# Praise for *Measuring the Immeasurable Mind*

"Matthew Owen's *Measuring the Immeasurable Mind: Where Contemporary Neuroscience Meets the Aristotelian Tradition* is a superb read, exemplary in scope and imagination. In addition to dismantling the philosophical foundations of physicalism, Owen presents a compelling case that dualists are in a suitable position to account for the neural correlates of consciousness. Skillfully deploying recent empirical evidence, Owen formulates a new model of the neural correlates of consciousness that coheres with Aristotle's insights regarding the formal character of biological systems and a dualist metaphysics of mind. A must-read for those who take a wide methodological scope in the philosophy of mind and neuroscience."

—**Eric LaRock**, Oakland University

"There is a growing discussion about what contribution hylomorphism can make to the philosophy of mind. Owen's book offers up answers to numerous issues and questions hylomorphists face. It will be of great interest to academics interested in the crossover between (Neo-)Aristotelian metaphysics and questions concerning consciousness, the mind-body problem, and the nature of neural correlation."

—**Nikk Effingham**, University of Birmingham

"*Measuring the Immeasurable Mind* is a unique book at the cutting edge of integrative philosophy of mind. Bringing together his considerable expertise in neuroscience and philosophy, Owen's bold offering shows that the recent findings in empirical science can be harmonized easily with a specific version of hylomorphism in an epistemically responsible way. The proffered harmonization makes clear how a robust dualist human ontology is fully consistent with the employment of the physical neural correlates of consciousness in attempting to quantify various states of consciousness. With its publication, no one who wants to be informed about recent, central developments in neuroscience and philosophy of mind can afford to neglect this work."

—**J.P. Moreland**, Biola University

"In *Measuring the Immeasurable Mind*, Owen develops what he names the Mind-Body Powers Model of Neural Correlates of Consciousness as a metaphysical resolution of the mind-body problem. This model cogently supports mind-body dualism against materialism based on what the author calls Neo-Thomistic hylomorphism, a concept developed from interpretations of ideas such as grounding and en-forming historically articulated by Aquinas and Aristotle. The incorporation of the older philosophical tradition with contemporary neuroscientific theories of consciousness offers a unique and important contribution to the philosophy of mind."

—**Michael L. Woodruff**, East Tennessee State University

"For theological anthropology to flourish, relevant work in neuroscience simply must be taken into account. Sometimes, however, theologians (and the religious communities they seek to serve) tend to worry that recent work in science will undermine traditional beliefs they take to be important, and, as a result, hold such advances at arm's length. In other cases, meanwhile, theologians are quick to reject traditional doctrine 'because science says so.' But what if such reactions are unwarranted and even unfortunate? This important work from Matthew Owen offers real help to the theologian, for he demonstrates that neither the science nor the theology need be threatened by the other. This book is well-informed and well-written; it is also both charitable and wise. It is a book that I will recommend eagerly and return to often!"

—**Thomas H. McCall**, Asbury University

"Owen's *Measuring the Immeasurable Mind* convincingly argues that the existence of the footprints of consciousness in the brain, the famed neural correlates of consciousness, is fully compatible with a dualistic view of the mind-body problem informed by Aristotle and Aquinas."

—**Christof Koch**, Allen Institute for Brain Science

"Owen touches a nerve in mainstream reductionism and physicalism. His ambitious attempt to connect solid metaphysics with the most advanced science of consciousness deserves extremely careful consideration. *Measuring the Immeasurable Mind* is a much-needed and refreshing view on the long-debated topics of mind-brain causation and the nature of conscious states. Some may not be convinced by his bold proposals, but everyone can learn a lot from his original examination that dusts off an ancient discussion in philosophy and science, revitalizing and updating Aristotelian and Thomistic hylomorphism. Overall, this is a rigorous, clear, and accessible book that makes an important contribution to its field."

—**Andrea Lavazza**, Centro Universitario Internazionale, Italy

"Owen's new book is fresh, bold, stimulating! It makes us rethink issues in contemporary neuroscience and in the history of philosophy alike, defamiliarized and thus put in a new light by the way the author makes them interact in this splendid book. Even those who might find themselves in disagreement with the methodology and/or the research results, will be stimulated by the arguments Owen puts forward and will want to engage in trying to find objections and counterarguments. This is a book that takes the reader through untrodden paths in a trailblazing way."

—**Anna Marmodoro**, Durham University

"Owen's *Measuring the Immeasurable Mind* brings a breath of fresh air to one of the most hotly debated issues of the nature of consciousness. Drawing insights from Aristotelian/Thomistic metaphysics, Owen presents a compelling argument that shows the metaphysical possibility of empirically discerning and quantifying irreducible consciousness. This is an excellent book which will be of great interest both for philosophers and neuroscientists who work on consciousness research. You cannot afford to bypass it."

—**Mihretu P. Guta**, Biola University

"It is rare these days to find authors doing research on the metaphysics of mind whose work exhibits ontological seriousness while engaging fruitfully with relevant scientific investigations on the mind. Owen's project in *Measuring the Immeasurable Mind* is an exception to the rule. Whether or not one finally agrees with Owen's conclusions or the metaphysical assumptions that guide his project, there is much to be gained by engaging with the account he offers and defends in this book. This book provides a model for how to do philosophical work in consciousness studies that amounts to more than mere conceptual policing. It is a must-read for anyone interested in the metaphysical foundations of the science of consciousness."

—**Andrei Buckareff**, Marist College

# Measuring the Immeasurable Mind

# Measuring the Immeasurable Mind

## Where Contemporary Neuroscience Meets the Aristotelian Tradition

Matthew Owen

LEXINGTON BOOKS
*Lanham • Boulder • New York • London*

Published by Lexington Books
An imprint of The Rowman & Littlefield Publishing Group, Inc.
4501 Forbes Boulevard, Suite 200, Lanham, Maryland 20706
www.rowman.com

6 Tinworth Street, London SE11 5AL, United Kingdom

Copyright © 2021 The Rowman & Littlefield Publishing Group, Inc.

Excerpts from Thomas Aquinas, *The Treatise on Human Nature*, translated by Robert Pasnau, Copyright © 2002 by Hackett Publishing Company, Inc. reprinted by permission of Hackett Publishing Company, Inc. All rights reserved.

*All rights reserved.* No part of this book may be reproduced in any form or by any electronic or mechanical means, including information storage and retrieval systems, without written permission from the publisher, except by a reviewer who may quote passages in a review.

British Library Cataloguing in Publication Information Available

**Library of Congress Cataloging-in-Publication Data**

Library of Congress Control Number: 2021933812

ISBN: 978-1-7936-4012-3 (cloth)
ISBN: 978-1-7936-4014-7 (pbk)
ISBN: 978-1-7936-4013-0 (electronic)

*To Aryn,
who brilliantly loves
the patients in her care.*

# Contents

| | |
|---|---|
| Preface | ix |
| Acknowledgments | xi |
| List of Common Abbreviations | xiii |
| **1** The Immeasurable Conscious Mind | 1 |
| **2** Neural Correlates of Consciousness | 17 |
| **3** Mental Causation: Identifying Dualism's Problem | 45 |
| **4** The Causal Pairing Problem | 73 |
| **5** Neo-Thomistic Hylomorphism | 87 |
| **6** En-forming Causal Pairing | 113 |
| **7** The Mind-Body Powers Model of NCC | 139 |
| **8** Empirically Discerning and Measuring Consciousness | 169 |
| References | 195 |
| Index | 213 |
| About the Author | 225 |

# Preface

Toward the end of the previous century, when physicalism was philosophical orthodoxy, the neuroscience of consciousness set its roots and sprouted. Within a materialist milieu, searching for the neural correlates of consciousness and a way to measure consciousness seems sensible.

If consciousness is reducible to or determined by physical neural correlates of consciousness (abbreviated NCC) which are measurable, it is reasonable to think that we can measure consciousness once we identify its neural mechanisms. But if physicalism falters and consciousness turns out to be irreducible and nonphysical, then such an optimistic endeavor may appear devoid of any realistic foundation. Thus, if dualism continues to be dusted off and reconsidered, as it has been in recent years, this might seem to threaten fundamental objectives in the science of consciousness. The worry is that the nonphysical conscious mind of dualism is an immeasurable problem for the science of consciousness, threatening the possibility of ever developing a practical means for empirically discerning and quantify consciousness.

Contrary to this concern, *Measuring the Immeasurable Mind* argues that the possibility of measuring consciousness, with its corresponding medical benefits, is not ruled out by a dualist human ontology that disavows the doctrines of physicalism. The author proposes the Mind-Body Powers model of NCC, which provides motivation for the contemporary search for NCC and a philosophical foundation for the possibility of empirically discerning and quantifying consciousness. The model gives a metaphysical explanation of NCC informed by Aristotle's understanding of causation and a substance dualist human ontology inspired by the thought of Thomas Aquinas, who often followed Aristotle. A practical implication is that consciousness researchers need not subscribe to physicalism to be rationally justified in their search for NCC and a practical means for empirically discerning and measuring consciousness. Physicalists *and* dualists can co-labor on one of the most practically pressing issues in consciousness research.

However, the aim of this work is a thoroughgoing waste of time *if*, as many claim, dualism is undermined by the mere existence of NCC and renders mental causation incoherent. Therefore, these concerns are addressed from the outset. After an introductory first chapter, dualism's dominant potential defeaters are explicated in the next three chapters. Chapter 2 clarifies the objection to dualism based on NCC and argues that NCC are neutral vis-à-vis physicalism and dualism. The following two chapters distinguish and elucidate dualism's principal problem regarding mental causation—the causal pairing problem. Chapter 5 introduces and explicates neo-Thomistic hylomorphism. Chapter 6 applies the view to the causal pairing problem, providing a solution that appeals to a fundamental tenet of hylomorphism. Then chapter 7 presents the Mind-Body Powers model of NCC, which is subsequently applied in chapter 8 to empirically discerning and quantifying consciousness.

*Measuring the Immeasurable Mind* addresses interrelated issues in the neurobiology of consciousness and philosophy of mind. Therefore, it's written for an interdisciplinary academic audience—from philosophers to physicists—interested in consciousness science. With precision and clarity an oft-forgotten, yet richly developed, historical vantage point is applied to contemporary consciousness research.

# Acknowledgments

This work is dedicated with love and utmost appreciation to my wife and dearest friend, Aryn Owen. I am indebted to her not only for her constructive feedback on my writing, but also for the relentless encouragement that she and my grandmother, Kathryn Owen, consistently provided as I worked on this monograph.

I am deeply grateful to Christof Koch not only for providing feedback on my work and serving as my fellowship advisor during my Tiny Blue Dot Fellowship, but also for years of mentoring me. Throughout this work, I have benefited from his expertise and insightful comments on earlier drafts. I also want to thank Elizabeth R. Koch and the Tiny Blue Dot Foundation for the honor of serving as the Elizabeth R. Koch Research Fellow for Tiny Blue Dot Consciousness Studies, from 2018 to 2020. The Tiny Blue Dot Foundation has not only made my research possible through their generous support but also enjoyable by providing a collegial interdisciplinary research team to be a part of. During our meetings, I have learned much from my TBD teammates, especially Giulio Tononi, Marcello Massimini, and Mélanie Boly. I also wish to thank Gonzaga University and my co-advisor Brian Clayton for their support during my research fellowship.

I have received helpful feedback on the ideas expressed in this work from many colleagues in various disciplines, from philosophy to physics. For helpful comments on chapters and related conference presentations or relevant insightful conversations, I would especially like to thank Nikk Effingham, Jussi Suikkanen, Anna Marmodoro, Henry Taylor, Giulio Tononi, Matteo Grasso, Erick Chastain, Francesco Ellia, Larissa Albantakis, Jonathan Lang, Nicco Reggente, Daniel D. De Haan, Dawid Potgieter, Gary Bartlett, Melia E. Bonomo, James Kintz, Cassie Finley, Jaden Anderson, Ross Inman, William M.R. Simpson, J.P. Moreland, Eric LaRock, and Mihretu Guta.

I also benefited from the University of Oxford's 2016 summer school *(Neo-)Aristotelian Approaches to the Metaphysics of Mind*, directed by Anna Marmodoro, Giulio Tononi's 2018 course on consciousness at the Neuroscience School of Advanced Studies in Venice, and the opportunity to visit the Center for Sleep and Consciousness at the University of Wisconsin-Madison during June of 2019. Each venue provided wonderful opportunities to learn from other researchers and to share ideas.

I am very grateful to Jana Hodges-Kluck and Sydney Wedbush for guiding this work through the publication process. As this process took place during a worldwide pandemic, thanks are also due to Nicolette Amstutz, Holly Buchannan, Hannah Fisher, and the editorial team as a whole at Lexington Books (Rowman & Littlefield) for their assistance in bringing this project to completion during difficult times. This work also benefited from feedback received during the review process. Anyone who is willing to serve as an anonymous book reviewer during a pandemic is an admirable servant of scholarship. I especially appreciate the insightful and detailed feedback I received from a particular anonymous reviewer on the second and eighth chapter which significantly strengthened the relevant sections.

Finally, I am grateful to my family, especially Aryn and our daughter Emma, as well as our dear friends who make life so enjoyable. Your support has been invaluable.

# List of Common Abbreviations

## WORKS OF THOMAS AQUINAS

| | |
|---|---|
| ST | *Summa Theologiae* |
| QDA | *Questions on the Soul* |
| SCG | *Summa Contra Gentiles* |
| QDSC | *Disputed Questions on Spiritual Creatures* |
| QDP | *On the Power of God* |
| *Sent.* | *Commentary on Peter Lombard's* Sentences |

## WORKS OF ARISTOTLE

| | |
|---|---|
| DA | *On the Soul* (or *De Anima*) |
| *Phys.* | *Physics* |
| *Meta.* | *Metaphysics* |

## NEUROBIOLOGY OF CONSCIOUSNESS

| | |
|---|---|
| NCC | neural correlates of consciousness |
| IIT | Integrated Information Theory |
| GWS | Global Neuronal Workspace Theory |

## DISORDERS OF CONSCIOUSNESS

| | |
|---|---|
| DOC | disorders of consciousness |

| | |
|---|---|
| VS | vegetative state |
| UWS | unresponsive wakefulness syndrome |
| MCS | minimally conscious state |

*Chapter 1*

# The Immeasurable Conscious Mind

> Students of physics and biology point out the necessity of measurement to science, and claim that as mental phenomena are not subject to measurement, psychology cannot become an exact science. They tell us only the brain may be studied by scientific methods. Students of philosophy, on the other hand, absorbed in the questions of speculative philosophy—What is the mind? What is its destiny and meaning?—are apt to look upon any attempt at measurement as trivial, if not profane.
>
> —J. McKeen Cattell, *Columbia College*

The above words are from "Mental Measurement," a research article written by one of the first psychology professors in the United States and the fourth president of the American Psychological Association. *The Philosophical Review* published the article in 1893. Whether it was anticipated by Professor Cattell or not, during the following century a plausibility structure would be widely embraced that suggests measuring the mind is theoretically possible. A year prior to the article's centennial anniversary, John Searle correctly noted at the outset of his widely influential book, *The Rediscovery of the Mind*, that materialism had become philosophical orthodoxy (Searle 1992, p. xii).

Materialism, simply put, claims all that exists is material. Physicalism can be seen as a view implied by materialism that focuses more precisely on the mind. While more nuanced details will be discussed in due course, for now it suffices to say that according to physicalism mental phenomena are physical phenomena, or at least depend on physical facts. On this view, physics is fundamental and fixes, or grounds, the mental. Whether mental states are reducible to physical states or not, they are at least tethered to and determined by their physical base.[1] Mental conscious states such as the qualitative sensations of feeling joy or sadness or vivid pain are present in virtue of the underlying neurobiology in one's cerebral cortex.

The physicalist framework provides a metaphysical foundation for the possibility of measuring our conscious mind. What we practically need to do to meet this objective, which is theoretically possible given physicalism, is map out which specific physical states are identical to, or at least underlie, specific mental states. This would allow us to empirically recognize the presence of mental phenomena like consciousness and measure it by measuring the underlying neurobiology. Since the physicalist framework provides a metaphysical basis for empirically detecting and measuring consciousness, it provides a philosophical foundation for a science of consciousness.

What if, however, physicalism falters?[2] Is there a foundation for a science of consciousness beyond (or perhaps after) physicalism? One alternative being considered is panpsychism, a view that claims consciousness is ubiquitous throughout the natural world.[3] Elsewhere, I have addressed panpsychism as it relates to the Integrated Information Theory of consciousness (see Owen 2019b, pp. 178–183). In this work, I do not take exception with panpsychism but simply offer a different alternative that is often neglected in contemporary philosophy of mind and cognitive neuroscience. I intend to demonstrate that the Aristotelian tradition provides a metaphysical foundation for a science of consciousness. I will argue that a model of neural correlates of consciousness informed by an Aristotelian understanding of causation and Thomas Aquinas's view of human nature provides a basis for the possibility of empirically detecting and measuring consciousness. What is especially peculiar about this basis is that it appeals to a substance dualist ontology of human persons that unabashedly violates the doctrines of physicalism.

In both its reductive and nonreductive forms, physicalism provides a metaphysical foundation for the theoretical possibility of empirically discerning and quantifying consciousness.[4] I intend to show that there is a substance dualist position that can do likewise.

My ultimate goal is not to settle a philosophical debate about the mind-body problem, nor is it to prove that consciousness is quantifiable and therefore my field deserves the respect physicists enjoy. In my opinion, neither the existence of consciousness nor the value of philosophy of mind depends on consciousness being quantifiable. My objective is merely to broaden the philosophical foundation for something very practical. In their article "Measuring Consciousness in Severely Damaged Brains," Olivia Gosseries, Haibo Di, Steven Laureys, and Mélanie Boly point out that the success of techniques utilizing advanced neuroimaging for accurately diagnosing disorders of consciousness is "intrinsically linked to understanding the relationship between consciousness and the brain" (Gosseries et al. 2014, p. 457). How consciousness relates to the brain affects what we can justifiably infer about whether unresponsive patients are conscious based on the neuronal processes in their brain. Thomism already has a voice in bioethics discussions relevant to disorders of consciousness such as coma and the vegetative state (see Shewmon 1997; Eberl 2006, Chs. 3 and 5).[5] I intend to apply a Thomistic-inspired view of human ontology specifically to the issue of how to interpret neurophysiological data when discerning whether an unresponsive patient is conscious.

Coupled with an Aristotelian metaphysics of causation, a view I call "neo-Thomistic hylomorphism" informs the Mind-Body Powers model of neural correlates of consciousness. This model, I will argue, provides a metaphysical foundation for the possibility of empirically discerning and measuring consciousness. Given that, empirically detecting and quantifying consciousness is possible not only within a physicalist framework, but also a nonphysicalist dualist framework. Since neo-Thomistic hylomorphism is a substance dualist position, an implication is that physicalists *and dualists* can co-labor in research focused on developing the means for empirically discerning and measuring consciousness in unresponsive patients. That is the ultimate upshot of this work. However, it is not until the final chapter that the foundation has been properly laid for me to address detecting and quantifying consciousness in unresponsive patients. Therefore, in order to clarify the overall trajectory and argument of the book, it will be helpful to further introduce the fundamental issue and the view I will apply to it, as well as key concepts and related topics.

## DETECTING CONSCIOUSNESS

When trying to discern the level of consciousness in patients, medical practitioners often ask questions, give commands, or provide sensory stimuli. To the degree a patient responds appropriately, you can reasonably infer that the patient has a corresponding level of consciousness. The responses can vary from verbal answers to questions that demand accessing memories to blinks of the eye as a form of "Morse code" communication with the outside world (see Massimini and Tononi 2018, pp. 34–38). In cases where motor movement is altogether impaired, a patient might be asked to imagine playing tennis so the expected corresponding neural activity in the motor cortex can be detected, indicating a neuronal response (see Owen et al. 2006).[6] There are some cases, however, when patients give absolutely no response whatsoever and yet they might be covertly conscious. In such cases it would be advantageous to have a way to empirically detect the presence of consciousness as well as quantify a patient's level of consciousness.

Yet, as suggested above, the possibility of doing so is related to the nature of consciousness and how it relates to the brain. If, as physicalism claims, consciousness is reducible to or determined by something physical that is empirically discernible and quantifiable, then consciousness is empirically discernible and quantifiable, at least in theory.

In her 1984 hit song *Material Girl*, Madonna summarized the prevailing ideology when the contemporary science of consciousness began setting its roots: "We are living in a material world and I am a material girl." The pop star echoes a description of materialism and its cousin physicalism in *The Cambridge Dictionary of Philosophy*: "Physicalism in the widest sense of the term, [is] materialism applied to the question of the nature of mind" (Shoemaker 1995, p. 617). Materialism claims that everything that exists is material, and physicalism applies this to the mind. So according to materialism/physicalism, Madonna is indeed a material girl with a material mind living in a material world. In the early 1990s, Searle simply stated the obvious in his aforementioned assessment of philosophy of mind: "Mainstream orthodoxy consists of various versions of 'materialism'" (Searle 1992, p. xii). At that time, Francis Crick and Christof Koch published their seminal article "Toward a Neurobiological Approach to Consciousness" (Crick and Koch 1990). What has followed is the contemporary search for neural correlates of consciousness, also known as NCC, the centerpiece of consciousness science concerned with measuring consciousness.[7]

The materialist milieu in the late twentieth century proved to be fertile ground for the rise of consciousness research in neuroscience, where it had long been taboo.[8] Whether it is acknowledged or not, explicit and implicit philosophical presuppositions can profoundly impact consciousness research.[9] While materialism and physicalism reigned supreme for the latter half of the twentieth century, and remain dominant today, the times are a-changin'—as the prophet, Bob Dylan, would say (see Koons and Bealer 2010; Göcke 2012). As suggested below in the section "Dusting Off Dualism," there are indicators that physicalism is floundering while dualism is being reconsidered. This might be seen as a threat to consciousness science, which set up shop in a physicalist factory, with the aim of detecting and measuring consciousness via its neural correlates.

If consciousness is reducible to or determined by its physical substrate that's empirically discernible and measurable, consciousness is empirically discernible and measurable. On the flip side, if consciousness is irreducible and nonphysical, and not necessarily determined by a physical substrate, one might worry that it cannot be empirically discerned nor quantified. The overarching aim of this work is to respond to this worry. I will argue that a nonphysicalist, substance dualist view regarding the ontology of consciousness does not necessarily entail the epistemological view that consciousness cannot be empirically discerned nor quantified.

My case will rely on the Mind-Body Powers model of NCC, which, I will argue, provides a basis for the possibility of detecting and measuring consciousness empirically. This model gives a metaphysical explanation of NCC that is informed by Aristotelian causation and neo-Thomistic hylomorphism, a substance dualist view of human ontology. If my overall conclusion is correct, researchers need not subscribe to physicalism to be rationally justified in their search for an empirical means of identifying covert consciousness in unresponsive patients. Even the substance dualist can work on this practically pressing issue in consciousness research.

I am, however, wasting my time *if*, as some claim, dualism is undermined by the mere existence of NCC or, as more claim, it renders mental causation incoherent. If either claim is true, the recent reconsiderations of dualism discussed below will only lead to a dead end in due course. Therefore, en route to the final chapter on detecting and measuring consciousness from the standpoint of a dualist human ontology, I will address the leading objections to dualism based on mental causation and neurobiology.

## INTERDISCIPLINARY INTEREST

Consciousness is of interest to researchers in a wide variety of disciplines, from philosophy to physics and almost everything in between (see Guta 2019a, pp. 2–3). As a philosopher who specializes in philosophy of mind and philosophy of cognitive neuroscience, I have written *as* a philosopher. Nonetheless I have written *for* neuroscientists, psychologists, and a variety of specialists fascinated by consciousness and its relationship to the body, who are willing to think hard about the relevant issues.

In the fourth century BC, Aristotle noted at the beginning of *On the Soul*, "To attain any knowledge about the soul is one of the most difficult things in the world" (Aristotle, DA 402a10-11).[10] Nearly two and a half millennia later, we are forced to agree by the subject matter itself—though we might reword Aristotle's statement in terms of consciousness. Yet, as Aristotle also notes in the same context, knowledge of the soul contributes profoundly to the advance of knowledge in general (see Aristotle, DA 402a5-6). The same might be said today of consciousness. By understanding it more fully, we can understand ourselves, and perhaps our place and purpose, more fully. Although the study is arduous, the trek is worth it.

A good place to begin is with clarifying the contemporary context. Because philosophical presuppositions often profoundly impact consciousness research, I will sketch a brief history of recent trends in philosophy of mind questioning physicalism and reconsidering dualism. The subsequent section will introduce dualism's potential defeaters and outline my approach to defeat such defeaters in the forthcoming chapters.

## DUSTING OFF DUALISM

Throughout the previous century, dualism's demise seemed imminent. During the 1970s, Daniel Dennett (1978, p. 252) assessed the field of philosophy of mind, and his evaluation of dualism was rather unflattering:

> Since it is widely granted these days that dualism is not a serious view to contend with, but rather a cliff over which to push one's opponents, a capsule "refutation" of dualism, to alert if not convince the uninitiated, is perhaps in order.

Dennett's (1978, p. 252) one-paragraph capsule refutation, which he went on to give, claims that a dualist has two bad options. She can accept an epiphenomenalism that claims there are causally impotent nonphysical mental states resulting from brain events or Cartesian interactionist dualism. Either way, Dennett (1978, p. 252) concludes, dualism comes with an "exorbitant price."

Two decades after Dennett's assessment, John Haldane (1998) published "A Return to Form in the Philosophy of Mind." In this article, written just before the turn of the century, Haldane (1998, p. 257) highlighted an ironic shift in the field: "Dualism has to be contended with." The two decades between the two assessments, Dennett's and Haldane's respectively, included the rising realization that physicalism faces its own troubles regarding mental causation and that consciousness is quite recalcitrant.

"What is new and surprising," wrote Kim (2005, p. 9), "about the current problem of mental causation is the fact that it has arisen out of the very heart of physicalism." According to the causal exclusion problem, the mental is allegedly excluded from doing any causal work because physical properties perform all the causal tasks in a physical world closed off from nonphysical causes (see Kim 2005, Chs. 1 and 2). Regarding the recalcitrance of consciousness, reducing consciousness to physics was made difficult by qualia and multiple realizability. Qualia are the subjective experiences of what-it's-like to be in particular conscious states. For example, there is a difference between consciously tasting chocolate versus consciously feeling a bee sting. This is a difference of qualia. Multiple realizability is the idea that conscious states like tasting chocolate can be realized by different physical states. Accordingly, two subjects with different types of physical bodies could have the same conscious state. Both qualia and multiple realizability will be discussed in chapter 2 (see the section "The Identity Theory's Viability").

To be sure, physicalism's troubles with mental causation and consciousness have not led to a resurgence of dualism, but a reconsideration.[11] It was true at the end of the previous century, and has been true during the beginning of this century, that physicalism is philosophical orthodoxy. Nevertheless, a change in the tide, a reconsideration of dualism, is evidenced by a brief overview of the publication record. The decades following Dennett's decisive declaration saw a variety of substantial publications questioning physicalism and supporting nonphysicalist and dualist views. Here I will give a brief overview of noteworthy works from 1980 to the present.

In the early 1980s, the Gifford Lectures given by the well-known neuroscientist and dualist, John Eccles, were published under the title *The Human Psyche*. In that volume Eccles (1980) argued against materialism and for a dualist-interactionist position. A couple years later, Howard Robinson (1982) published *Matter and Sense: A Critique of Contemporary Materialism* with Cambridge University Press (hereafter CUP). The same year, Frank Jackson (1982, p. 130) presented his well-known Knowledge Argument against physicalism using his infamous thought experiment involving a hypothetical neuroscientist named "Mary" (see also Jackson 1986).

In the mid-1980s, Oxford University Press (hereafter OUP) published *The Evolution of the Soul* in which longtime Oxford professor Richard Swinburne (1986) argues for substance dualism. Two years later, W. D. Hart (1988) argued for substance dualism in *The Engines of the Soul*, published by CUP. Hart was well aware that he was swimming against the current of mainstream materialism. "But orthodoxy needs devil's advocates," he claimed for the following reason: "They have a serious part in the play of ideas even if committed heterodoxy invites excommunication" (1988, p. x). Before the decade closed, the University of Virginia Press published the edited volume, *The Case for Dualism* (Smythies and Beloff 1989).

The 1990s also saw substantive publications that questioned physicalism as well as publications that favored dualism. The year of 1991 witnessed several publications. Oxford Press published David Hodgson's (1991) *The Mind Matters: Consciousness and Choice in a Quantum World*, which challenged a reductive physicalist mechanistic view of the mind. Roderick Chisholm published "On the Simplicity of the Soul" in *Philosophical Perspectives*. Chisholm (1991, p. 167) argued that the nature of human persons is completely unlike that of compound physical things. Routledge published John Foster's (1991) *The Immaterial Self: A defense of the Cartesian dualist conception of the mind*. Two years after the prolific year of 1991, a volume edited by Howard Robinson (1993) entitled *Objections to Physicalism* was published by OUP.

Evidently, Hart was not excommunicated for defending his unorthodox dualist position in the 1980s. In the mid-1990s, he wrote a section on dualism for *A Companion to Philosophy of Mind*, published by Blackwell. Also in the mid-1990s, CUP published E. J. Lowe's (1996) *Subjects of Experience*, in which Lowe defends a distinctly non-Cartesian version of dualism. And OUP published David Chalmers's (1996) influential work *The Conscious Mind*, in which Chalmers endorses a form of property dualism. Just before the turn of the millennium, William Hasker (1999) advocated for emergent substance dualism in *The Emergent Self*, published by Cornell University Press.

The new millennium brought new works challenging physicalism as well as new works supporting dualism. In 2001, CUP published *Physicalism and Its Discontents*. In 2005, David Oderberg published "Hylemorphic Dualism" in *Personal Identity*, a volume published by CUP and edited by Ellen Frankel Paul and colleagues. Brie Gertler first published "In Defense of Mind-Body Dualism" in 2008, which was republished several years later in *Consciousness and the Mind-Body Problem*, published by OUP. *The Waning of Materialism*, edited by Robert Koons and George Bealer, was published by OUP in 2010. This volume includes intriguing sections authored by leading philosophers, such as Laurence BonJour's chapter "Against Materialism" and E. J. Lowe's chapter "Substance Dualism: A Non-Cartesian Approach."

In 2011, Continuum published *The Soul Hypothesis*, which includes contributions by Dean Zimmerman, William Hasker, and Daniel Robinson. The same year, in the article "No Pairing Problem," published in *Philosophical Studies*, Andrew Bailey and company contend that Jaegwon Kim's argument against substance dualism based on the causal pairing problem fails. The following year witnessed the publication of *After Physicalism*, which includes chapters by the likes of E. J. Lowe, Alvin Plantinga, John Foster, Richard Swinburne, and Howard Robinson.[12] The same year, the journal *Philosophia Christi* published "A New Argument Against Materialism" by Plantinga and "An Empirical Case Against Central State Materialism" by Eric LaRock.

Also in 2012, Christof Koch, who co-instigated the contemporary search for neural correlates of consciousness with Francis Crick, diverged from today's dominant doctrine. In *Consciousness: Confessions of a Romantic Reductionist*, Koch (2012, p. 152) writes:

> The dominant intellectual position of our day and age is physicalism—at rock bottom all is reducible to physics. There is no need to appeal to anything but space, time, matter, and energy. Physicalism—a halftone away from materialism—is attractive because of its metaphysical sparseness. It makes no additional assumptions. In contrast, such simplicity can also be viewed as poverty. This book makes the argument that physicalism by itself is too impoverished to explain the origin of mind. In the previous chapter, I sketched an alternative account that augments physicalism. It is a form of property dualism: The theory

of integrated information postulates that conscious, phenomenal experience is distinct from its underlying physical carrier.

To some, Koch's willingness to diverge from the physicalism so widely assumed in neuroscience might be surprising. For the historian of neuroscience, however, Koch's intellectual courage might be reminiscent of Wilder Penfield's (1975, p. 114) words: "as a scientist, I reject the concept that one must be either a monist or a dualist because that suggests a 'closed mind.'"

In 2013, Laurence Bonjour published "What Is It Like to Be Human (Instead of a Bat)?" in the *American Philosophical Quarterly*. He argues that a physical, neurophysiological account of a human person is "radically incomplete" as it lacks many significant facts about one's conscious mental life and thus "physicalism or materialism, as an account of human beings, is surely and irredeemably false" (BonJour 2013, p. 383). The year of 2013 also saw the arrival of two monographs in favor of dualism. Richard Fumerton's (2013) *Knowledge, Thought, and the Case for Dualism* appeared in the Cambridge Studies in Philosophy series, and Richard Swinburne's (2013) *Mind, Brain, & Free Will* was published by OUP. One year later, *Contemporary Dualism: A Defense* was added to the Routledge Studies in Contemporary Philosophy series. Tomas Bogardus also argued that prima facie justification for dualism persists undefeated in the aptly entitled article "Undefeated Dualism," which appeared in the journal *Philosophical Studies*.

In the international philosophy journal *Teorema*, I argued in 2015 that we can know a priori that physicalism and a priori knowledge are epistemologically incompatible; hence, being a member of philosophical orthodoxy in the philosophy of mind comes at a high price. The year 2015 also witnessed the start of a study funded by the Templeton Foundation. *The New Directions in the Study of the Mind*, which was based at the University of Cambridge and led by Tim Crane, investigated nonphysicalist views of the mind. In 2016, CUP published Howard Robinson's *From the Knowledge Arguments to Mental Substance: Resurrecting the Mind*, in which he makes a case for substance dualism.

Dualism doubled down in 2018 with the début of *The Blackwell Companion to Substance Dualism*. According to Michael Tye's endorsement, "This excellent book is chock-a-block full of interesting essays." Its thirty-two chapters include contributions for and against various versions of dualism and its physicalist rivals, written by a lineup of heavy hitters, including: E. J. Lowe, Richard Swinburne, J. P. Moreland, Robert C. Koons, Joshua Rasmussen, Jaegwon Kim, Peter van Inwagen, Lynne Rudder Baker, and Tim Bayne. The same year, Brandon Rickabaugh published "The Primacy of the Mental: From Russellian Monism to Substance Dualism" in *Philosophia Christi*. It is also worth noting that Howard

Robinson, J. P. Moreland, Angus Menuge, and Charles Taliaferro contributed to *Christian Physicalism: Philosophical Theological Criticisms* published by Lexington Books. This edited volume offers objections to physicalism based in philosophy, theology, and spiritual formation. In the journal *Medicine, Health Care and Philosophy*, Hane Htut Maung (2018) published an article responding to the claim in psychiatric literature that neuroscientific research regarding the biological basis of mental disorders undermines dualism in philosophy of mind. Before the year of 2018 closed, the Central European University hosted a conference entitled *Dualism in the Twenty-First Century*.

In 2019, Routledge published *Consciousness and the Ontology of Properties*, edited by Mihretu Guta. In this volume, J. P. Moreland argues that substance dualism provides the best explanation for the unity of consciousness and Richard Swinburne argues against the "supposed causal closure of physics," which is often considered an essential doctrine of physicalism. I, too, make a contribution, arguing that the simplicity argument for physicalism and against dualism based on NCC is found wanting. In a research article published in *Synthese*, I argued that if the Integrated Information Theory of consciousness embraced dualism, it could circumnavigate the causal pairing problem by following hylomorphism's example laid out in chapter 6 of the present work. Also in 2019, Richard Swinburne made a case for substance dualism that appeals to personal identity in *Are We Bodies or Souls?* published by OUP.

Physicalism is still the zeitgeist of our day, and dualism is far from dominant. That said, it is apparent that dualism is being reconsidered and revamped, which has led to the reconsideration and development of considerable arguments for dualism. Nevertheless, if dualism is to be considered as anything more than an interesting route to a dead end, its potential defeaters must be dealt with.

## DEFEATING DUALISM'S DEFEATERS

The third chapter of Jaegwon Kim's *Physicalism, Or Something Near Enough* is entitled "The Rejection of Immaterial Minds: A Causal Argument." Kim's (2005, p. 70) opening words allude to dualism's reconsideration:

> The deep difficulties that beset contemporary nonreductive physicalism might prompt some of us to explore nonphysicalist alternatives; in fact, the nonreductivist's predicament seems to have injected new vigor into the dualist projects of philosophers with antecedent antiphysicalist sympathies.

In the first two chapters of the book, Kim focuses on problems physicalists face regarding mental causation and consciousness. According to Kim (2005, p. 70), the upshot is that physicalists have two options: epiphenomenalism or reductionism. "With good reason," Kim writes, "most philosophers have found neither choice palatable." Epiphenomenalism renders the mental causally impotent and thus strikes many of us as "obviously wrong," as it smacks against what we know about ourselves as agents; reductionism seems to many "not much better" since the mental wouldn't provide any meaningful addition to physical processes doing all the causal work (Kim 2005, pp. 70–71).

Given the two unpalatable options, the worry is that substance dualism will be considered as an alternative (Kim 2005, p. 71). But dualism is a dead end that won't help and will actually make things worse when it comes to mental causation, argues Kim (2005, p. 71). As the title of the chapter suggests, we should reject substance dualism based on its inadequacy regarding mental causation. Throughout the chapter, Kim (2005, pp. 71, 73–74, 92) tries to defeat dualism by developing a famous objection Princess Elisabeth of Bohemia raised against René Descartes's interactionist dualism. The result is the causal pairing problem for substance dualism, which I will explain in detail in chapter 4. The basic idea is that a nonphysical mind and a physical body are of such different natures that it is impossible for them to be causally paired as cause and effect. The purported consequence of the causal pairing problem is that mental causation is impossible for the nonphysical (or immaterial) minds of substance dualism. If that's true, dualism should be rejected.

If the causal pairing problem is irremediable, it could itself undermine dualism. Although it is not dualism's only prominent potential defeater. A second considerable objection appeals to neuroscience. Hedda Hassel Mørch (2017) sums up a common conception: "Modern science has given us good reason to believe that our consciousness is rooted in the physics and chemistry of the brain, as opposed to anything immaterial or transcendental." This claim is so widely believed today that it can be (and often is) made without citing any of the evidence that allegedly supports it. Nonetheless, it is allegedly neural correlates of consciousness that provide the powerful evidence that suggests some version of physicalism is true (see, e.g., Murphy 1998, p. 13). If neuroscience proves physicalism, then it disproves dualism, which then becomes an "antiscientific" position in contradistinction to the "scientific" physicalist alternatives (cf. Searle 1992, pp. 3–4). Related to this second potential defeater, though distinct from it, is the practical concern discussed above (see the section "Detecting Consciousness"). That is, if consciousness is nonphysical and not grounded in something physical, then it cannot be empirically recognized and measured.

Despite reconsiderations of dualism in recent times, substance dualism faces these potential defeaters that threaten its viability. I intend to defeat these potential defeaters. I will advocate for a version of substance dualism I call *neo-Thomistic hylomorphism* and demonstrate how it overcomes dualism's potential defeaters. This dualist view of human nature, it will be argued, provides a basis for causal pairing between a mental cause and its physical bodily effects. And together with principles gleaned from Aristotle's view of causation, it provides a basis for a good explanation of neural correlates of consciousness, which grounds the possibility of consciousness being empirically discernible and quantifiable.

The next three chapters explicate dualism's potential defeaters. The following four chapters explain neo-Thomistic hylomorphism and apply it to the pairing problem, NCC, and measuring consciousness. Chapter 2 will focus on neural correlates of consciousness, explaining what they are, how we identify them, what they entail, and the argument against dualism based on such correlations. The conclusion of the chapter is that neural correlates do not entail any particular view of the mind, and the argument against dualism based on these correlations has considerable weaknesses.

Chapters 3 and 4 clarify substance dualism's paramount problem regarding mental causation. There are various problems pertaining to mental causation, and not all of them are serious problems for dualism (cf. Kim 2001b, pp. 271–272). Therefore, these chapters focus on clarifying what problem is dualism's paramount problem regarding mental causation and thoroughly explaining it. Chapter 3 will distinguish two problems that do *not* provide strong threats to substance dualism from the causal pairing problem, which does. I will discuss a problem Donald Davidson raises regarding a lack of psychophysical laws and the causal exclusion problem. My intention is to distinguish these problems from substance dualism's chief problem regarding mental causation—the causal pairing problem—and justify why I set them aside to focus on the pairing problem. Chapter 4 explains the pairing problem in depth.

Chapter 5 introduces neo-Thomistic hylomorphism. My intent in this chapter is to present and describe a view in the philosophy of mind and human ontology that is informed by Thomas Aquinas. However, my focus will not be exegetical or historical. I will not focus on interpretive debates about what exactly Aquinas thought. I call my view a *"neo-*Thomistic" view because it is informed by Aquinas's thought, but I make no claim that it is exactly what Aquinas thought. I am interested in whether this view can defeat substance dualism's defeaters. I think it can, and that is what I intend to demonstrate.

Since I am advocating for a view that I consider a substance dualist position, by arguing that it can overcome dualism's primary objections, it will be helpful to clarify what I mean by "substance dualism." According to property dualism, there are nonphysical mental properties even if the mind itself is a physical substance. According to substance dualism, the mind is a nonphysical substance.[13] Property dualists say there are nonphysical mental properties.

Substance dualists go further and say such mental properties are the properties of a nonphysical mental substance (cf. Robinson 2020, section 2.3). What I mean by "substance dualism" is the idea that human persons have, or are, a nonphysical mental substance.

Admittedly, different metaphysical commitments entail different definitions of a substance. I will discuss what a substance is according to Aristotelian-Thomistic metaphysics at the beginning of chapter 5 before presenting neo-Thomistic hylomorphism. For now, let me simply say that a substance is a unified entity that can persist through time and change. Therefore, what I mean more precisely by "substance dualism" is broadly the view that human persons have, or are, a unified nonphysical mental entity that can persist through time and change.

In contemporary philosophy, the term "mind" usually refers to a mental substance, whether that substance is physical or nonphysical. Hence, physicalists will speak of a physical mental substance that they call the "mind." Similarly, substance dualists often speak of a nonphysical mental substance that they call an "immaterial mind" or "nonphysical mind," following the terminology of modern philosophy. I will often do the same. However, substance dualists also sometimes refer to the nonphysical mental substance using the term "soul," following premodern terminology. I will often use this terminology as well, and especially when describing Aquinas's thoughts or the neo-Thomistic view I'm advocating.

It's worth explicitly mentioning that, contrary to what's often thought, not all substance dualists think that a human person consists of two substances. There are substance dualists that claim the mind (or soul) is a nonphysical substance *and* the body is a physical *substance*. But there are also substance dualists that merely say that the mind is a substance without saying that the body is itself a substance. For example, a crude caricature of substance dualism presents the picture of a ghost, who we can call Casper, in a machine that Casper inhabits and somehow moves around. Suppose someone held this Casper dualist view but her metaphysical view of substances only allowed her to say Casper is a substance, but not the machine body. She would still be a substance dualist because substance dualism is not about counting two complete substances. Rather, substance dualism is *dualist* in the sense that it claims there are nonphysical mental properties, and it is *substance* dualist in the sense that it says there is a nonphysical substance that bears the nonphysical mental properties.

One more nuance is worth noting. Just as there are property dualists that claim the mind is a physical substance that bears nonphysical properties, there are substance dualists that claim the mind is a nonphysical substance that has physical properties. In his chapter on dualism in *The Oxford Handbook of Philosophy of Mind*, E. J. Lowe clarifies that according to his non-Cartesian substance dualist view, a human subject is not reducible to their body nor any physical part of it, but is nevertheless the bearer of physical properties in virtue of having a physical body (Lowe 2009, p. 68). As will become apparent, the neo-Thomistic hylomorphic view I propose also permits the claim that the nonphysical soul has physical properties of the body in a derivative sense (see chapter 5).

After explicating neo-Thomistic hylomorphism, I will apply the view to substance dualism's paramount problems in the final three chapters. The sixth chapter will present a solution to the causal pairing problem that capitalizes on the fundamental tenet of neo-Thomistic hylomorphism. The seventh chapter will introduce an Aristotelian powers ontology and apply it to the proposed hylomorphic view in order to give an account of neural correlates of consciousness (i.e., NCC). My aim is to present a model informed by Aristotelian causation and neo-Thomistic hylomorphism—the Mind-Body Powers model of NCC—which provides a good explanation of NCC. The model is unique in that it's informed by a substance dualist view of human ontology—neo-Thomistic hylomorphism—which diverges from a physicalist framework. Chapter 8 addresses the possibility of empirically discerning and measuring consciousness from the standpoint of the Mind-Body Powers model. There I will argue that the empirical discernibility and measurability of consciousness is not necessarily ruled out by a nonphysicalist, dualist human ontology.

To be clear, this book will *not* provide sufficient justification for substance dualism broadly, nor neo-Thomistic hylomorphism specifically. My goal is not offensive in the sense of giving a positive case for dualism or lethal objections to physicalism. Rather, my goal is defensive in the sense that I am responding to substance dualism's strongest objections and a practical concern related to diagnosing disorders of consciousness. Simply put, I intend to demonstrate that neo-Thomistic hylomorphism coupled with Aristotelian causation can defeat dualism's paramount potential defeaters and provide a rational basis for the scientific study of consciousness. Given that, it is a position worthy of serious consideration.

## PRESUPPOSITIONS

We like to think that as rational individuals we can approach a topic in an entirely objective manner utterly unadulterated by our own preliminary ideas. Consequently, it is not as common as it ought to be for starting assumptions or presuppositions to be explicitly stated. Yet, whether one is a philosopher, a scientist, or simply a thoughtful thinker trying to be as objective as possible, we all have starting presuppositions with which we inevitably approach any research topic with. In fact, our presuppositions often inform the very topics we choose to study.[14]

We must allow the facts to revise our starting assumptions and presuppositions; nevertheless, we approach any intellectual inquiry with our presuppositions intact, even if they will be later revised. It is important to acknowledge not only this but also our presuppositions themselves. In addition to being a matter of honesty, it is a matter of being rationally thorough, since the validity and epistemic role of unacknowledged presuppositions is more likely to go unexamined. When presuppositions are explicitly acknowledged, readers can more easily examine such starting points and be aware of the epistemic role they play in various lines of reasoning.[15]

In light of my defensive objective, my starting presuppositions are especially important. I find the recent reconsiderations of dualism warranted because I think there are substantive objections to physicalism.[16] Furthermore, some arguments for dualism broadly and substance dualism more specifically strike me as persuasive and sufficient to justify dualism.[17] I find persuasive Dean Zimmerman's (2011) vagueness argument from property dualism to substance dualism based on the subject that is the bearer of experiences. I also think William Hasker's (2010) argument from the unity of consciousness justifies substance dualism.[18] E. J. Lowe's (2001) argument from the simplicity of the self is especially compelling. Although he favors a Cartesian view that differs from my Thomistic-inspired view, Richard Swinburne's (1986) overall case for substance dualism in *The Evolution of the Soul*, especially the evidence from personal identity, is quite strong in my opinion. I also think there are good arguments for substance dualism based on freewill and human knowledge (cf. Swinburne 2013). In short, I think there are good arguments that demonstrate physicalism is false and some version of dualism is correct. Unlike many of my physicalist colleagues, this is the vantage point from which I begin. However, I also think there are objections to dualism that must (and can) be dealt with. Hence the *telos* of the following pages.

## NOTES

1. Cf. Kim (2011, pp. 12–13; 2005, Ch. 6).
2. See Koons and Bealer (2010) and Göcke (2012).
3. See Koch (2012, Ch. 8) and Goff (2019); cf. Tononi and Koch (2015, p. 11) and Koch (2019b, Ch. 14). Goff et al. (2017) provide an introduction to panpsychism.
4. I use "detect/discern" as well as "measure/quantify" synonymously.
5. As discussed in chapter 8, the vegetative state is more fittingly called unresponsive wakefulness syndrome.

6. See chapter 8 for a discussion of the paradigms used to discern a patient's level of consciousness.

7. Cf. Frith and Rees (2017); Hohwy and Bayne (2015, pp. 161–163); Kandel and Hudspeth (2013, p. 18); Overgaard (2017); Storm et al. (2017).

8. Baars (2003) discusses the fading of the taboo.

9. Regarding the influence of philosophical positions on neuroscience research related to consciousness, see Bennett and Hacker (2003), Shulman (2013), and Owen and Guta (2019).

10. Aristotle's *On the Soul* is also referred to as *De Anima*, which is often abbreviated DA.

11. Compare to Timothy O'Conner and David Robb's (2003, p. 5) assessment: "Dualism in recent years has even enjoyed something of a renaissance."

12. Furthermore, in 2012 a relevant interdisciplinary research project funded by the Templeton Foundation entitled "Neuroscience and the Soul" kicked off at Biola University (see Crisp et al. 2016).

13. Timothy O'Conner and David Robb (2003, p. 4) likewise distinguish substance dualism and property dualism.

14. On the epistemic role of presuppositions in cognitive neuroscience research, see Koch (2004, p. 21).

15. On how this applies to theoretical approaches to consciousness research, see M. Owen and Guta (2019, pp. 12–13).

16. Elsewhere, I have argued against physicalist tenets (see M. Owen 2015, 2020d).

17. For an introductory overview of arguments for dualism, see Robinson (2020) and Lowe (2009). For in-depth works presenting arguments for various versions of dualism, see Chalmers (1996), Lavazza and Robinson (2014), and Loose et al. (2018).

18. However, I don't think the unity of consciousness supports Hasker's version of emergent substance dualism as well as it supports other versions of substance dualism.

*Chapter 2*

# Neural Correlates of Consciousness

The nonphysical mind of substance dualism is commonly considered a relic of our pre-neuroscientific past. Neuroscience, many would say, has nullified dualism. According to philosopher Nancy Murphy (1998, p. 13), neuroscience provides "dramatic evidence for physicalism." In *The Astonishing Hypothesis: The Scientific Search for the Soul*, molecular biologist and neuroscientist Francis Crick (1995, p. 3) proposes that human persons are "in fact no more than the behavior of a vast assembly of nerve cells and their associated molecules."[1] Even theologian Michael Horton (2011, p. 376) informs the faithful in *The Christian Faith: A Systematic Theology for Pilgrims on the Way* that neuroscience has firmly established that "the mind is matter (i.e. the brain)."[2]

Granted, not everyone shares such sentiments. There are numerous dissenters. Nonetheless, most would still agree with a concession found in the *Stanford Encyclopedia of Philosophy*: "The 'neuroscientific milieu' of the past four decades has made it harder for philosophers to adopt dualism" (Bickle et al. 2010). True as that may be, we must ask: What is the evidence modern neuroscience provides against dualism? Surely it is not merely the fact that acts of the mind such as thought are associated with the brain. For this is hardly new information provided by modern neuroscience. The Greek physician Hippocrates (460–375 BC) knew this much, as did medieval philosopher and theologian, Thomas Aquinas (see Hippocrates 1886, p. 344; Aquinas, ST 1a 78.4c). So what original evidence has modern neuroscience provided that disproves dualism?

The foremost answer is: neural correlates of consciousness (for brevity NCC). Accurate or not, the most common example of an NCC is C-fiber activation in one's brain that takes place when they are in a mental state of pain.[3] Accordingly, when I am in a conscious state of pain, there's a corresponding neural state in my brain, that is, C-fiber activation. The C-fiber activation is the neural correlate of the conscious state of pain, according to this example.

On the basis of such correlations, it is argued that physicalism is true and, thus, dualism is false. This chapter is devoted to explicating and analyzing this line of reasoning. In the first section, I will clarify what NCC are. Then the methods used to identify NCC will be introduced in the following section, where I will briefly summarize some example studies. The third section will focus on what NCC imply. The fourth section will explicate and critically analyze the argument against dualism based on NCC and the principle of simplicity.

In short, this chapter will clarify the objection to dualism that's based on NCC and highlight its shortcomings in order to cast considerable doubt on its sufficiency to undermine dualism. In chapter 7, I will then argue that neo-Thomistic hylomorphism is not only consistent with NCC but also provides a good explanation of NCC. It should be noted that throughout this entire work I won't concern myself with the question of whether there are NCC (cf. Noë and Thompson 2004). I assume there are. My concern is what they imply, or what best explains them.

## WHAT IS AN NCC?

Simply put, a neural correlate of consciousness is a neural state or process that correlates with consciousness. The idea is that when a subject is in a particular conscious state, there is a corresponding state of their brain that is correlated with their conscious state. That's a basic description of an NCC.

However, in his influential work "What Is a Neural Correlate of Consciousness?" David Chalmers (2000) points out that there are various conceptions of neural correlates within the NCC literature. Thus, he tries to offer conceptual clarity by giving a theoretically neutral, reasonable definition that reflects common usage (Chalmers 2000, pp. 31, 38). As a starting point, Chalmers (2000, pp. 17–18) presents and considers a definition of an NCC derived from a conference program for the Association for the Scientific Study of Consciousness:

> A neural system N is an NCC if the state of N correlates directly with states of consciousness.

Chalmers (2000, p. 18) then asks two clarifying questions: (1) What are the relevant states of consciousness? (2) What does it mean for a neural state(s) to correlate directly with states of consciousness?

Regarding the first question, Chalmers (2000, pp. 18–23) surveys several classes of phenomenal consciousness sometimes considered in the NCC literature. The first class is being conscious. A relevant NCC would be a neural state that correlates with a subject being conscious versus not being conscious. The second is a background state of consciousness—such as being awake, asleep, under hypnosis, in a state of flow, or the like. A corresponding neural correlate would be a neural state that directly correlates with one being under hypnosis.

The third class Chalmers (2000, p. 19) covers is contents of consciousness. Suppose that after a day of teaching I come home from campus with roses. My wife will delightfully rush over and smell them. When she does, her experience of the smell of the roses is a specific content of her consciousness. A neural state that correlates with that particular experience of the smell of the roses would be a relevant NCC. The final class Chalmers (2000, p. 22) considers is arbitrary phenomenal properties. Specific states of any of the above classes can be members of this class, which might be useful if one tries to give a general definition of an NCC.

Before moving on to consider Chalmers's second question, I want to make a terminological point that will become relevant in chapter 8. Neuroscientists often make a distinction between what they call the *full NCC* versus *content-specific NCC*. The full NCC corresponds to one's overall conscious state such as being conscious versus unconscious; content-specific NCC correspond to conscious states with specific content such as a color or a face (Boly et al. 2017, p. 9604; Koch et al. 2016, p. 308; Storm et al. 2017).[4]

When it comes to Chalmers's second question, the relevant complexity becomes quite apparent. The original question is: What does it mean for a neural state to directly correlate with states of consciousness? Yet this question prompts Chalmers (2000, pp. 24–28) to ask two more fundamental questions. First, must the neural state be necessary, sufficient, or necessary and sufficient for the conscious state it's correlated with?[5] Second, must the correlation hold across all cases or only across specific types of cases (i.e., cases with ordinary brain function in an ordinary environment, cases with a normal brain but unusual inputs, cases with varying stimulation, or cases with abnormal brain function due to lesions)?

Adequately answering the above questions is beyond the scope of this section. My aim is not to formulate an original definition of an NCC. I intend to demonstrate that a substance dualist position, neo-Thomistic hylomorphism, provides a good account of NCC. Such an account will be most effective if it assumes a reasonable definition that accords with general usage and is theoretically neutral. Since Chalmers's (2000, pp. 31, 38) definition is intended to meet such objectives, I'll adopt his definition:

> An NCC is a minimal neural system N such that there is a mapping from states of N to states of consciousness, where a given state of N is sufficient, under conditions C, for the corresponding state of consciousness.[6]

While Chalmers's definition is more precise than the one from the conference program, there are several parts that need explaining. First, the phrase "*minimal* neural system" needs clarification. Here Chalmers (2000, p. 24) is trying to avoid irrelevant neural processes being included (cf. Koch et al. 2016, p. 308). To clarify this point, let's consider a "minimal engine system" that is a correlate of a car starting. The car starts when the ignition switch turns. The turning of the ignition switch is a correlate of the car starting. However, there are other conditions true of the car when it starts. The gas tank will contain gas. The fuel line will be clear. The spark plugs will be clean. The crankshaft will be in place. The list goes on and on. Yet, if I were to explain to a new driver which of the above is a relevant correlate pertaining to their car starting, I need not explain the entire system of a properly functioning car engine. I only need to tell the new driver about the "minimal engine system" that's a correlate of the car starting—that is, when the ignition switch turns, the car starts.

Likewise, when it comes to an NCC, we are not concerned with everything taking place in the entire nervous system, or even the brain in particular, when one is in a particular conscious state. Rather, we're concerned with the minimal neural state(s) or process(es) that correspond to that conscious state. To refer again to the common example regarding the mental state of pain and C-fiber activation, suppose my dog Anselm accidentally bites my hand while playing. Is every synapse between the bite on my hand and my brain a neural correlate of my pain state? No; rather, the neural correlate is the minimal neural system that, under certain conditions, is sufficient for me to feel the corresponding state of pain. According to our example, that would be the corresponding C-fiber activation, which is the minimal neural system since there is no more fundamental system that suffices for the corresponding state of pain (cf. Chalmers 2000, p. 25). This is just one hypothetical example. Each minimal neural system will vary depending on the conscious state it correlates with.

The second part of Chalmers's definition that needs explanation is the qualifier "under conditions C." To return to our car analogy, while the ignition switch turning is the correlate of the car starting, there are further conditions true of the engine when the car starts. As mentioned above, the gas tank will have gas, the fuel line will be clear, the spark plugs clean, and so on. Such conditions are conditions of a normally functioning car engine. According to Chalmers (2000, p. 31), the conditions typically relevant to NCC include normal brain functioning that permits some atypical inputs and brain stimulation but not changes to brain structure (e.g., lesions).[7]

Lastly, let's clarify the phrase "there is a mapping from states of N to states of consciousness." First of all, this is not meant to suggest that there's only an NCC if we have already identified it and mapped it. Rather, there is a mapping from the neural state $N^1$ to the conscious state $C^1$ if the former corresponds with the latter so that the correspondence could be mapped if identified. Secondly, the idea of mapping between corresponding neural states and conscious states pertains to subjects across a species, not just an individual subject. However, this mapping across a species is not necessarily a correspondence of identical neural states in every subject in a particular conscious state. The search for NCC is a search for biological regularities, not necessary identical correspondence relations. Biological regularities of all kinds permit variations.

For elucidation, let us again return to the familiar example of C-fibers and pain. Regarding the human species, there's a mapping from C-fiber activation to the conscious state of pain if it's true that when humans experience pain, their conscious experience corresponds with C-fiber activation in their central nervous system. This doesn't, however, rule out variation. After all, pain can be one aspect of someone's overall conscious experience that includes additional mental states, which might result in neuronal variations. For example, some endurance athletes mentally train themselves to have an unusually high tolerance for pain through self-talk that affects their overall conscious experience when they're in pain. Thus, while their experience of pain will be similar to the experiences of other subjects in pain, there will be some mental variation that could result in variation with respect to the neural processes.

Such variation is also relevant to a methodological challenge regarding controlled experiments. That is, it is very difficult, if not impossible, to produce identical overall conscious experiences in subjects being tested. If a neuroscientist showed me an image of red roses, I could report to her my conscious perception of the image, just as my wife could if shown the same image. Yet, my overall conscious experience will likely vary from hers even though we both consciously perceive the same image. After all, I do not care much for red roses, whereas she absolutely loves red roses and gets very excited about them. Yet, even if the neuroscientist could get us to have the exact same overall conscious experience, our brains are not exactly similar. In fact, no two individuals have brains that are exactly alike, not even identical twins, or even clones. In light of such variations, we should not expect the search for NCC to reveal correlations that are exactly the same across a species, but rather similar correlations reflecting biological regularities that permit variation.[8]

In sum, according to Chalmers's definition, an NCC is a minimal neural system that's sufficient under certain conditions for the corresponding state of consciousness, such that this correspondence can be mapped. Before concluding this section, it's worth noting that Chalmers's definition can be modified to apply specifically to certain types of phenomenal consciousness. For example, Chalmers (2000, p. 31) gives a modified definition particularly relevant to contents of consciousness:

> An NCC (for content) is a minimal neural representational system N such that representation of a content in N is sufficient, under conditions C, for representation of that content in consciousness.

While this modified definition applies to contents of consciousness, similar modifications could be made so that the definition applies specifically to other types of phenomenal consciousness. At times, such modifications might even be necessary.

Now that we are equipped with a definition of neural correlates of consciousness, let's consider how we identify NCC. The next section introduces standard methodology used to identify neural correlates of consciousness.

## IDENTIFYING NCC

The aim of this section is to introduce methods used to identify neural correlates of consciousness. In meeting this objective I will refer to several example studies. The first study took place in the nineteenth century and provided a theoretical basis for techniques vital to the contemporary search for NCC. The second study, published in the year 2000, pertains to neural correlates of binocular rivalry. It gives us an example of standard contemporary methodology used to identify NCC, which explicitly relies on subjective reports from study participants. The third study also pertains to neural correlates of binocular rivalry, but it implements the recently developed "no-report paradigm." This paradigm includes trials of the study with explicit reports from participants, as well as trials without explicit reports where physiological measures (e.g., pupil dilation) are used to infer what participants perceive (Koch et al. 2016, p. 308).

Fundamental to identifying neural correlates of consciousness is finding neural activity that consistently corresponds with certain conscious states. Imaging brain activity is central to this endeavor. Over a century ago, Italian physiologist Angelo Mosso (1846–1910) laid the conceptual basis for brain imaging techniques vital to the contemporary search for NCC (see Sandrone et al. 2014). A chief challenge to studying the brain is that it's enclosed in a hard protective casing—the skull. Mosso worked with a patient named Bertone, who suffered extensive damage to the top of his skull; consequently, much of it was missing. Where Bertone's skull was missing, Mosso placed a cap made out of a rubberlike substance, *gutta percha* (Glickstein 2014, p. 343). This flexible cap made it possible to record brain pulsations of blood pressure correlated with mental activity such as emotional arousal and doing arithmetic (Glickstein 2014, p. 343).

Mosso's study confirmed a straightforward hypothesis. That is, if the brain works harder, there will be increased blood flow to the brain, so if the brain works harder when the mind works harder, there will be increased blood flow to the brain when the mind works harder. Put differently, (a) increased mental activity means (b) increased brain activity, which means (c) increased blood flow in the brain. So (a) increased mental activity correlates with (c) increased blood flow in the brain. Mosso confirmed this by measuring the increased blood flow in Bertone's brain that took place when Bertone's mental activity increased.

However, the method Mosso used to measure pulsations in Bertone's brain had a limitation. It was effective only if the patient had an abnormal skull breach (Sandrone et al. 2014, p. 622). Mosso's ingenious "human circulation balance" was invented to overcome this limitation (Sandrone et al. 2014, p. 622). Mosso had his patients lay on a table that was essentially a balance intended to measure pulsations of blood flow that would tip the balance. Whether or not Mosso's human circulation balance was reliable, his work laid the conceptual foundation for noninvasive functional brain imaging techniques (Sandrone et al. 2014, pp. 621–622). Noninvasive functional brain imaging is done while the brain is active and without being invasive to the brain by penetrating it in any significant way. Such brain imaging is vital to the search for NCC insofar as it allows us to detect brain activity corresponding with a conscious subject's mental activity.

Needless to say, noninvasive functional brain imaging technology has advanced significantly since Mosso's day. Positron emission tomography (PET) provides an example, even though its use has dropped off since the 1980s, and functional magnetic resonance imaging (fMRI) provides another. A PET scan can reveal blood flow, glucose metabolism, oxygen metabolism, or concentrations of dopamine transporter indicative of brain activity. An fMRI reveals hemodynamic activity (i.e., blood flow and volume) indicative of brain activity. Blood carries oxygen on molecules of hemoglobin containing iron that changes the thermodynamic and magnetic properties of the brain area when there is an increase of blood flow corresponding to increased activity. This is detected by magnetic resonance imaging. In addition to modern developments in noninvasive brain imaging, in the 1920s a German psychiatrist, Hans Berger, discovered that it is possible to record electrical activity in the brain from the human scalp (Glickstein 2014, p. 338). This type of recording is called an electroencephalogram, or EEG, which is commonly used today.

These technological advances have made possible research methods employed in the contemporary search for NCC, principally instigated in the late twentieth century by Francis Crick and Christof Koch (1990). To elucidate the method of inquiry in NCC research, which relies on such technology, let's consider a conventional study led by Alex Polonsky to identify neural correlates of perception during binocular rivalry (Polonsky et al. 2000). The phenomenon of binocular rivalry was discovered in the nineteenth century by the English scientist Sir Charles Wheatstone (1838).[9] Today binocular rivalry is commonly used in NCC research.[10] It takes place when both eyes are presented simultaneously with different stimuli, and the brain allows only one stimulus to be consciously perceived at a time. The perceived stimulus is often referred to as the dominant stimulus, while the one that is not perceived is the suppressed stimulus.

In normal circumstances, both eyes are presented with the same physical stimulus (e.g., both of my eyes are presented with my computer screen as I write). Nevertheless, there is a subtle disparity between the information received by each retina. Yet, we do not have two subtly different perceptions, one for each eye. For example, I perceive the bookshelf to the left of my desk which my left eye is slightly closer to than my right eye, but I don't perceive the spatial difference. Rather than having one percept (via my left eye) of the bookshelf being 20 inches away and a second (via my right eye) of the bookshelf being 21.5 inches away, I just have a single percept with the information received by each eye fused into a single conscious perception that allows me to know how far away the bookshelf is. The brain puts the information from both eyes together into one packaged perception.

Binocular rivalry takes place when each eye is presented with two different stimuli instead of the same stimulus. For example, the left eye might be presented with the image of a house, while the right eye is presented with the image of a face. In such cases, as Wheatstone was surprised to discover, only one stimulus is dominantly perceived by the subject, while the other stimulus is suppressed. Instead of the information received by each eye being fused into one percept, the subject perceives one or the other image at a time. For a brief period of time, they will see the image of the house while the image of the face is suppressed. Then they will see the image of the face for a brief period while the image of the house is suppressed. This phenomenon of binocular rivalry is important to NCC research because the information the brain receives remains constant while the visual conscious perception differs. This raises the question of whether there are specific neuronal mechanisms corresponding to the coming and going of each conscious perception.

In Polonsky's study, fMRI was used to measure fluctuations of cortical activity correlated with alternating perceptions during binocular rivalry. Two different rival stimuli were simultaneously presented to a particular eye of study participants as the researchers measured fMRI signals in the early visual cortex (Polonsky et al. 2000, p. 1153). One stimulus was a higher-contrast green grating image, presented to one eye. The second was a lower-contrast red grating image, presented to the other eye. The subjects would report which stimulus they perceived by pressing one button when they perceived the green image and another button when they perceived the red image (Polonsky et al. 2000, p. 1154). The researchers found that the fMRI signal in the primary visual cortex, V1, correlated with the perceptions the subjects reported. When the subjects reported seeing the higher-contrast green grating image, activity in VI tended to increase, and it tended to decrease when the subjects reported seeing the lower-contrast red grating image (Polonsky et al. 2000, p. 1155).

To clarify the degree of these activity fluctuations during binocular rivalry, the researchers did a separate series of fMRI scans to measure V1 activity. During this series, they presented the subjects with only one image at a time and alternated between the images so that there was no rival stimulus simultaneously presented. Then they compared the V1 activity during binocular rivalry to the V1 activity while there was only one stimulus alternately presented at a time. Confirming earlier accounts, it was found that neural activity in V1 followed the alternations of the stimuli (see Heeger et al. 2000, Polonsky et al. 2000, p. 1155). The researchers also found that the fluctuations of V1 activity during binocular rivalry were 45 to 83 percent as large as fluctuations in V1 activity induced during the stimulus alternations when only one image was presented at a time. Furthermore, V1 activity fluctuations were about equal to those in visual areas nearby—that is V2, V3, V4v, and V3a (Polonsky et al. 2000, p. 1155). Polonsky and his colleagues concluded that their findings suggest that neuronal activity corresponding to binocular rivalry is expressed as early as the primary visual cortex, V1 (Polonsky et al., 2000, p. 1157). As they acknowledge, this runs contrary to the view that such neural activity occurs predominantly in later visual areas, which prior studies indicated (Polonsky et al. 2000, p. 1153; cf. Leopold and Logothetis 1996; Logothetis and Schall 1989; Sheinberg and Logothetis 1997).

In short, the methodology used consisted of controlling the stimuli presented to each eye of the subjects and observing neuronal activity in the visual cortex using fMRI as the subjects reported their conscious perceptions. Such methodology critically depends upon the subject's awareness of their perceptions and the subject's report of what they perceive to the researchers. The researchers infer that the conscious perception being reported correlates with the neural activity at that time, which is identified via brain imaging. At this point, there is a methodological concern that researchers might actually be identifying the NCC of the mental activity related to the report rather than the conscious perception reported.[11]

When a subject perceives a green grating image without considering the fact that she perceives the image, she is in a different mental state than when she perceives the image *and* considers her perception of it as she reports her perception. In the former case, she simply has a first-order awareness of what's perceived (i.e., the green image). When she reports the perception, she has a second-order awareness of her first-order awareness and an intention to report her perception. Consequently, her mental state then includes perceiving the green image and considering the fact that she perceives the green image and her intention to give the report. The worry is that the identified neural correlate may actually be correlated with the mental state that includes the second-order awareness and intention when it is thought to be correlated with just the first-order mental state of simply perceiving the green image.

This worry motivates what is called the "no-report paradigm" (see Tsuchiya et al. 2015). This methodological paradigm includes study trials with explicit reports from participants as well as trials without explicit reports where physiological measures are used to infer what participants perceive (Koch et al. 2016, p. 308). In 2014, Stefan Frässle and a team of researchers did a study on binocular rivalry with a pioneering application of a no-report paradigm (see Koch et al. 2016, p. 318, note 36). The subjects were presented with different stimuli to induce binocular rivalry. Their no-report paradigm used the ocular motor reflex, optokinetic nystagmus, along with pupil size as "objective measures" of which stimulus was dominant (Frässle et al. 2014, p. 1739). These reflex measures together with fMRI purportedly allowed them to assess the NCC of binocular rivalry without active reports from the subjects (Frässle et al. 2014, p. 1738). The researchers compared these measures to trials with active reports from the subjects, which allowed them to test the applicability of the objective measures (Frässle et al. 2014, p. 1743). They concluded that active report and possibly introspection were partly responsible for neuronal activation patterns that are typically observed (Frässle et al. 2014, pp. 1743–1745). The no-report paradigm is an innovative approach, but its accuracy depends on the reliability of the objective measures, which may be difficult to establish.

This concludes our brief overview of the methodology used to identify NCC, in which we have briefly considered three example studies intended to identify NCC. Given the discovery of neural correlates of consciousness, it's fitting to ask what they imply.

## IMPLICATIONS OF NCC

While the evidence for neural correlates of consciousness may be clear, it is not obvious what such correlations entail regarding the nature of the mind.

At this point a common fallacy—*post hoc, ergo propter hoc*—is tempting to commit. This fallacy is committed when one infers that ϕ caused φ simply because ϕ is correlated with φ. If one inferred that Barack Obama winning the democratic nomination in 2008 caused the stock market crash of 2008 simply because there's a correlation between the two events, they would commit this fallacy. Correlation does not entail causation. One needs further rationale to infer that a correlation is best explained by (or suggests) a causal relation.

Likewise, a correlation by itself does not entail dependency, identity, or that one correlate is reducible to the other. Given that ϕ is correlated with φ, we need more information to justifiably conclude that ϕ depends on φ, is identical to φ, or is reducible to φ. Suppose that all around the world whenever any philosopher heard a knock on their door, they found a packaged philosophy book on their doorstep. On the basis of this correlation alone, we could not justifiably infer that the packaged books caused the knocks, lest we commit the fallacy mentioned above. Likewise, we could not infer that the door knocks depend on the philosophy books. Nor could we infer that the knocks are identical to the books or in some way reducible to them. Further information would be required to justifiably make such inferences.

In some cases where we have identified an NCC, we might have additional data that justifies further inferences. For example, suppose that whenever Fern was in a mental state of remembering her childhood in Kansas, a particular part of her brain was activated. Suppose further that the same area is activated in Kathryn's brain whenever she remembers her childhood in Washington State. Moreover, presume the data related to Fern and Kathryn confirms numerous studies with many human subjects. Given this, we could know that Fern and Kathryn's mental states of remembering their childhood correlate with neural activity in a particular part of the human brain.

From this correlation alone, we could not infer that Fern and Kathryn's mental state caused the neural activity, depended on it, was identical to it, or reducible to it. However, suppose that Fern and Kathryn lost the part of their brain with the neural correlates and directly after this they could never again remember their childhoods. The data set would then include more than just the correlations. With the additional data, it would be justifiable to conclude that their mental state of remembering their childhood was not only correlated with the neural activity, but also depended on it.

If a physicalist assumed physics is fundamental before mapping the correlation between Fern and Kathryn's mental activity and the corresponding neural activity, she would likely conclude that the correlation implies that the mental activity depends on the neural activity before gaining the additional data. She would be justified in doing so to the degree that her assumption was well justified. But she would not be arriving at this conclusion merely on the basis of the correlation. Rather, she would be justifiably arriving at the conclusion on the basis of the correlation *coupled with her preexperimental assumption*. Similarly, a dualist could justifiably conclude that Fern and Kathryn's immaterial minds stood in some type of causal relation with the neural correlates, if she justifiably assumed that their mental states are not reducible to physical states. However, the dualist, like the physicalist, would be arriving at her conclusion on the basis of the NCC and her preexperimental assumption.

Pessimists might roll their eyes at this point and remind us that everyone has pre-experimental assumptions. That's true. Nevertheless, that does not mean we cannot learn anything from NCC. It does mean, however, that it is important to analyze our preexperimental assumptions and to be aware of the justificatory role they play. According to Chalmers (1998, p. 227), "Once we recognize the central role of preexperimental assumptions in the search for the NCC, we realize that there are limitations on just what we can expect this search to tell us." Given this, it's fitting that Thomas Metzinger (2000, p. 4) writes the following in the introduction to his influential volume *Neural Correlates of Consciousness*:

> However, mapping does not mean reduction. Correlation does not mean explanation. Once strict, fine-grained correlations between brain states and conscious states have been established, a number of theoretical options are still open. Additional constraints therefore will eventually be needed. Important questions are What is the true nature of these psychophysical correlations? Are we justified in interpreting them as causal relations? What additional constraints would have to be introduced in order to speak of law-like correlations . . . ? Is a fully reductive account, or even an eliminativist strategy, possible?

Though Metzinger is no dualist (cf. Metzinger 2003), he goes on to acknowledge:

> Assume that we find a strict and systematic correlation between a certain brain property or type of neural event N and the subjectively experienced phenomenal property of "sogginess" S. This is entirely compatible with Cartesian dualism: The underlying relation could indeed be a causal one, namely causal interaction between events in two ontologically distinct domains.

In short, correlations don't entail causation, dependency, identity, or reducibility. Further philosophical argumentation beyond the empirical data of NCC is required to arrive at a justified conclusion about the nature of the mind.

## AGAINST DUALISM

So far, I have argued that NCC do not entail a particular view of consciousness. As Jakob Hohwy (2007, p. 461) puts it: "The notion of 'correlation' doesn't by itself commit one to any particular metaphysical view about the relation between (neural) matter and consciousness." Given this, a philosophical argument is needed to show that NCC suggest dualism is false.

In *Sensations and Brain Processes*, J. J. C. Smart (1959) aims to provide such an argument against dualism. Smart's influential argument is based on the principle of simplicity, also known as Occam's razor. Simplicity says that when there are multiple theories that sufficiently explain a data set, the theory that includes the least unnecessary entities is preferable. Simplicity is one theoretical virtue that helps us discern which theory among competing theories provides the best explanation.

In this section, I will present Smart's simplicity argument and then give a twofold critical analysis. After presenting the simplicity argument, I'll critique Smart's conclusion by summarizing problems that have convinced many philosophers that it is false. Then I'll highlight a handful of weaknesses regarding the argument itself. My focus will be Smart's simplicity argument, but it should be noted that he is not the only materialist who argues on the basis of simplicity. More recent contributions in this area have come from Christopher Hill (1991) as well as Ned Block and Robert Stalnaker (1999).[12] However, it is important to note that Hill's (1991, Ch. 2) simplicity argument against dualism markedly differs from Smart's (1959) argument. I find it best to focus on Smart's argument, rather than Hill's, for two reasons. First, Smart has been widely influential, and, second, Hill's argument and conclusion are much weaker, in my opinion.

According to Hill (1991, pp. 28–29), the simplicity argument can be formulated in three different ways according to different ways of understanding simplicity. Hill (1991, pp. 29–39) argues that two of the formulations fail. He then presents his own formulation, which relies on simplicity for aesthetic appeal rather than epistemic justification (Hill 1991, pp. 39–40). Hill (1991, p. 40) concludes:

> It seems, then, that my claim for the simplicity argument must be modest. I must not maintain that it can be used to establish that type materialism is probable, nor that it can be used to convert all rational beings to type materialism. Rather, I can claim only that the argument makes a case that will be found persuasive by people whose aesthetic intuitions cause them to attach importance to ontological simplicity. It is, of course, my hope that the reader will find on reflection that he or she belongs to this group.

There is much to be said for not overplaying your hand, but Hill's argument and conclusion seem *too* modest to seriously threaten dualism. Unlike Hill, Smart argues via an appeal to simplicity that it is indeed probable that materialism is true and dualism is false. If his argument is cogent, the implications for dualism are clearly consequential. Henceforth, I will focus on Smart's simplicity argument.

## Simplicity Argument

Appealing to simplicity, Smart argued for the thesis that sensations are identical to neural processes in the mid-twentieth century. Due to advances in brain and neural imaging technology, we now know more about the brain and how it corresponds to consciousness at the neuronal level. While NCC data has developed since Smart published *Sensations and Brain Processes* in 1959, his argument presupposes the core concept that neural processes correlate with conscious sensations.

Smart concluded that mental sensations are nothing more than their neural correlates. So pain, for example, is allegedly C-fiber activation according to an "is" of identity (Smart 1959, p. 145). The mental state of pain is nothing over and above the neural process of C-fiber activation. "They" are identical and thus have all and only the same properties.[13] This thesis is referred to as Smart's brain process theory or Smart's identity theory. Specifically, it is a type identity theory since it says sensation types (e.g., pain and tasting chocolate) are identical to types of neural processes. By contrast, a token identity theory says particular instances of a sensation (e.g., the pain I now feel after stubbing my toe) are identical to particular neural processes (cf. Smart 2007).

Smart (1959, p. 142) acknowledges that consciousness appears irreducible. But "for various reasons," writes Smart (1959, p. 142), "I just cannot believe that this can be so." For one, it seems to him "frankly unbelievable" that everything could be explicable in terms of physics and biology *except* for dualism's irreducible conscious states (Smart 1959, p. 142). Smart refers to such states as "nomological danglers" because they don't fit in a completely physical world. To make matters worse, these danglers need laws linking them to the brain processes they are correlated with (Smart 1959, pp. 143, 156).

Dualism allegedly includes these copious irreducible conscious states—nomological danglers—and the laws "whereby the 'nomological danglers' dangle" (Smart 1959, p. 156). The alternative is the identity theory, which does not include such odd ontological baggage. According to the identity theory, there are only brain processes that turn out to be sensations upon empirical investigation. Therefore, the identity theory is simpler and thus enjoys a theoretical virtue dualism allegedly lacks. Simply stated, simplicity favors the identity theory over dualism. Consequently, the identity theory is the best explanation of NCC data, according to the argument, which we can reconstruct as follows:

| | |
|---|---|
| (SIMPLICITY) | All else being equal, the simplest explanation of a data set is the best explanation. |
| (NCC-DATA) | Neuroscience has discovered NCC. |
| (ID-SIMPLER) | Relative to dualism, the identity theory is the simplest explanation of NCC. |

*Therefore:*

| | |
|---|---|
| (ID-BEST EX) | Relative to dualism, the identity theory is the best explanation of NCC. |

The first premise (SIMPLICITY) is assumed by everyone. The second premise (NCC-DATA) is likewise a safe assumption. The fact that there are NCC is the data to be explained. The third premise (ID-SIMPLER) is a key step in arriving at the conclusion. Dualism supposedly includes nonphysical mental states (i.e., "nomological danglers") and the relevant laws relating them to neural events, whereas the identity theory does not.[14] So the identity theory is a simpler theory than dualism. Hence the conclusion that the identity theory is the best explanation of NCC data, and therefore it is probably true. That is the simplicity argument; and now that it has been explained, we can analyze its merit.

The following subsection will focus on the conclusion (i.e., Smart's identity theory) and provide reasons to think it is false. After discussing the viability of the conclusion, I will critically analyze the argument for it.

## The Identity Theory's Viability

The simplicity of the identity theory is an attractive virtue. However, in his critical analysis of Smart's argument from simplicity, Jaegwon Kim (2011, p. 102) points out that what one person finds to be a simple explanation might seem like an inadequate, truncated explanation to someone else. In some cases, simplicity can be seen as "poverty," to echo Koch (2012, p. 152), who has recently argued that physicalism in general is "too impoverished" to explain the origin of consciousness. Critiques of physicalism generally, and the identity theory specifically, regarding explanatory inadequacy are not uncommon, and I will raise similar issues in the next section.[15]

However, given the truth of the identity claim central to the identity theory, expecting certain explanations may be unwarranted. After all, the truth of the identity theory would limit the need for some explanations. Consider the following analogy offered by Ned Block and Robert Stalnaker (1999, p. 24):

> Suppose one group of historians of the distant future studies Mark Twain and another studies Samuel Clemens. They happen to sit at the same table at a

meeting of the American Historical Association. A briefcase falls open, a list of the events in the life of Mark Twain tumbles out and is picked up by a student of the life of Samuel Clemens. "My Lord," he says, "the events in the life of Mark Twain are exactly the same as the events in the life of Samuel Clemens. What could explain this amazing coincidence?" The answer, someone observes, is that Mark Twain = Samuel Clemens. Note that it makes sense to ask for an explanation of the correlation between the two sets of events. But it does not make the same kind of sense to ask for an explanation of the identity. Identities don't have explanations. . . . The role of identities is to disallow some questions and allow others.

The reasoning here is sensible, and it's clear how it analogously applies to NCC. The events in the history of conscious state $C$ are exactly the same as those of neural state $N$, and, therefore, $C = N$; and this identity relation eliminates the need for certain explanations.

The identity relation itself does not need an explanation, and it eliminates the need to answer questions like "how can $C$ fit into a world full of physical things like $N$" and "how did $C$ emerge from $N$." Given that $C$ is identical to $N$, such questions are nonsensical, and therefore need no answer. Given the identity relation central to the identity theory, its advocate can appeal to this point. When allegations are made that the identity theory is explanatorily inadequate because it fails to explain something like the emergence of consciousness, the identity theorist can claim that such an explanation is unnecessary given the identity relation central to the identity theory. However, the strength of such a claim hinges on the truth of the identity theory.

The aim of this section is to question the truth of the identity claim foundational to the identity theory and to highlight reasons to think it is false. While the reasons I will provide are not incontestable, they are merely intended to provide warrant for thinking it is probable that the identity theory is false and it is worthwhile to look for alternatives. The reasons I offer come in the form of apparent differences between consciousness and its neural correlates. Consciousness appears to be different from its neural correlates in several respects. If it's probable that any one of these apparent differences is an actual difference, then there is reason to think that consciousness is not identical to NCC and therefore Smart's identity theory is false. The falsity of the theory would render its simplicity moot. Even if it cannot be shown that the apparent differences are indeed true differences, they nevertheless call into question the alleged identity relation central to the identity theory.

I will focus here on apparent differences between consciousness and neural correlates related to well-known and widely discussed issues in contemporary philosophy of mind: multiple realizability, the knowledge argument, epistemic access, and the possibility of zombies. These topics merit a book-length treatment. But since a significant amount of attention has been given to them elsewhere and they are not the focus of this work, I will only be highlighting key points and their present applicability.[16]

It is important to emphasize that my target in this section is *not* the explanatory adequacy of the identity theory per se. The same points I rely on can be used to argue for the explanatory inadequacy of the identity theory. However, this section focuses on the truth of the identity claim central to the identity theory. I'll use the same points someone else might use to argue that the theory is explanatorily inadequate. But I am arguing specifically that there are reasons to think the theory is false. Granted, if (A) the theory is false, then (B) it's explanatorily inadequate, as well. Nevertheless, my focus in this section is on giving reasons specifically in support of (A).

According to Leibniz's law of the indiscernibility of identicals, if $R$ is identical to $S$, then everything true of $R$ is true of $S$ and what is not true of $R$ is not true of $S$.[17] The foregoing discussion assumes this law and addresses apparent differences between consciousness and neural correlates, providing reasons to doubt the truth of the alleged identity relation central to the identity theory.

*Multiple Realizability*

More than any other difficulty the identity theory faces, multiple realizability can be accredited with bringing about its "unexpectedly early decline," according to Kim (2011, p. 122). The problem of multiple realizability, famously raised by Hilary Putnam (1967), consists of the idea that it is possible for mental kinds to be realized by multiple physical kinds. This possibility does not seem permitted by Smart's type identity theory.[18]

According to his identity theory, a conscious state is its neural correlate. If this were true, the conscious state could not exist without its neural correlate, which suggests that beings without the neural correlate could not have the conscious state. However, it seems that beings with different neurobiology than that of humans could have some of the same conscious states with different physical correlates. The common examples provided often include the idea that pain is correlated with C-fiber activation and point out that it's possible for there to be animals without C-fibers capable of experiencing pain. In addition, it's often thought that there could be aliens without C-fibers capable of experiencing pain. Neither such animal nor alien pain would be possible if pain is identical to C-fiber activation.

Multiple realizability is relevant to the contemporary neuroscientific study of NCC in two ways that provide some support to the idea that multiple realizability is possible. The full NCC of consciousness (i.e., the neural correlate of being conscious as opposed to NCC of specific conscious states) is not yet identified. Yet, the Integrated Information Theory of consciousness (for brevity IIT) is a prominent theory pertaining to the full NCC (see Tononi et al. 2016).[19] On IIT, multiple realizability is possible (cf. Fallon 2016; Tononi and Koch 2015). Hence, it is not surprising that a leading proponent of IIT, Christof Koch (2012, p. 152), accepts a form of property dualism, which entails a denial of the identity theory.

Secondly, multiple realizability is relevant to neuroplasticity. In the last two decades, there has been much research interest in neuroplasticity, which is the nervous system's capacity "to respond to intrinsic or extrinsic stimuli by reorganizing its structure, function and connections" (Cramer et al. 2011, p. 1592). In "What is a neural correlate of consciousness," Chalmers (2000, p. 24) ruled out the idea of an NCC being necessary and sufficient on the grounds that such a condition would be too strong. "It might turn out," Chalmers reasoned, "that there is more than one neural correlate of a given conscious state." It now appears that there is some empirical support suggesting that after traumatic injury resulting in the loss of NCC, consciousness can come to be correlated with different neural mechanisms than before (Jensen and Overgaard 2011; Mogensen 2011; Munoz-Cespedes et al. 2005; Overgaard and Mogensen 2011). If such alterations do take place, this suggests that multiple realization is not just a possibility but an actual occurrence in (some) brains affected by traumatic injury. Yet, the idea that this is even a possibility worth investigating suggests something important for our purposes. That is, there is an apparent difference between consciousness and its corresponding neural correlates—namely, consciousness is multiply realizable.

*Knowledge of Qualia*

In his article "Epiphenomenal Qualia," Frank Jackson (1982, p. 130) provides an infamous thought experiment pertaining to qualia, which are the felt experiences of what-it's-like to be in a conscious state. The experience of tasting Swiss chocolate feels different than the experience of tasting dirt, which is a difference of qualia. In other words, what-it's-like to taste Swiss chocolate is different than what-it's-like to taste dirt. Here I will offer a modified version of Jackson's thought experiment that also pertains to direct epistemic access to one's conscious states.[20]

Imagine a neuroscientist, Mary, whose parents want her to become the star of a famous thought experiment. From birth they have her wear black-and-white goggles, which she never takes off. As a result, she has never seen the color red. One day Mary gets a large grant to study all the neurobiological, neurophysiological, neurochemical information and every other type of physical information pertaining to the nervous system that corresponds to seeing red. If her study is successful, she will learn all the physical information there is about the NCC of seeing red.

Since the grant is very lucrative, Mary decides to hire the band members of the *Red Hot Chili Peppers* to be subjects in her study. Mary studies all that takes place in the brains of the band members as they view visual images that include the color red, which Mary presents them with. However, Mary cannot see the color in the images herself since she always keeps her goggles on (to stay true to her parents' dream). In the end, Mary's study is a major success and she learns all the physical information about the neural correlates of the conscious state of seeing red. However, she does not share any of this information with the band. For she knows the drummer is upset with the lead singer and is therefore leaving the band to pursue a career in neuroscience. She doesn't want him to publish the information from the study before her, so she keeps it to herself. However, despite her knowledge of all the physical information about the NCC of seeing red, it seems the band members would have knowledge Mary would lack. They would know what-it's-like to see red, whereas Mary would not because she has never seen red due to her black-and-white goggles.

Despite her knowledge of all the physical information about the NCC of seeing red, it's reasonable to think Mary would not know what-it's-like to see red. Yet, the band would know what-it's-like to see red, even though they don't know the physical information about the NCC. This suggests two differences. First, the physical information describes the neural correlates but not the corresponding qualia. Secondly, and perhaps more importantly, *qualia can be directly known* by the subject without knowledge of the physical information describing the neural correlates which *cannot be known in the same way* (see Swinburne 2013, Ch. 3). It is due to these differences that neuroscientists rely on reports from subjects when identifying NCC.[21] The difference of epistemic access is also why a theory of consciousness like IIT can reasonably be based on starting axioms about consciousness that are directly known (see Tononi et al. 2016).[22]

*Zombie Nervous Systems*

The possibility of zombies can also make one worry about the identity theory.[23] David Chalmers (1996, p. 94) describes a zombie as something physically the same as a conscious being such as himself, but with no conscious experience. If the identity theory is true, zombies are not possible. Given the identity theory, conscious states are identical to their neural correlates, and therefore neural correlates could not possibly exist without their corresponding conscious states. If zombies are possible, it would imply that the identity theory is false. Many philosophers think zombies are possible, and many do not. Nonetheless, to call into question the identity claim central to the identity theory, we only need the mere possibility of an unconscious nervous system, not a complete, embodied, behaving zombie. Thus, cerebral organoids, popularly known as "mini-brains," provide an interesting conversation piece to the debate.

In 2011, at the Institute of Molecular Biotechnology in Vienna, a postdoctoral researcher, Madeline Lancaster, inadvertently brought about the production of a brain organoid from human embryonic stem cells (Willyard 2015, p. 520). The brain organoids neuroscientists are now capable of growing consists of several million neurons. In April 2018, *Nature* published an article on the ethics of experimenting with these cerebral organoids. The team of authors, led by Nita A. Farahany (2018, p. 430), offers the following description of the organoids:

> Brain organoids can be produced much as other 3D multicellular structures resembling eye, gut, liver, kidney and other human tissues have been built. By adding appropriate signaling factors, aggregates of pluripotent stem cells (which have the ability to develop into any cell type) can differentiate and self-organize into structures that resemble certain regions of the human brain.

These so-called "mini-brains" reportedly resemble human brains in noteworthy ways regarding their constitution, neural activity, and structure.[24] Hence, they prompt a key question with serious ethical implications: Are they conscious?[25] The question is natural to ask because it seems possible that despite cerebral organoids being composed of human brain tissue, and having similar structural features, and neural activity, it's possible they are not conscious. To use Chalmers's terminology, it is possible that they are "zombie" mini-brains (see M. Owen 2020b).

Cerebral organoids are minuscule and far less developed than actual brains. Nonetheless, we can imagine a hypothetical scenario in which scientists are capable of growing a full-sized central nervous system that they can take parts from to replace damaged neural tissue in human patients. Even when we "life-size" these brain organoids, it would seem that the same question would still make sense: Are they conscious? This question would be far from trivial, given the ethical risks of experimenting with conscious subjects on one side and the potential medical benefits on the other.

The identity theory appears to rule out the possibility that such nervous systems with the same neural function could be nonconscious "zombie" nervous systems. But it seems possible that they could be. This suggests that it is possible for the neural correlates of conscious states to exist without the corresponding consciousness, which is obviously impossible for the conscious states. This appears to be another difference between neural correlates and corresponding conscious states.

In summary, in this section I have now offered several apparent differences between consciousness and neural correlates that suggest conscious states are not identical to NCC, and vice versa.[26] Given that consciousness is not identical to its neural correlates, the identity claim central to the identity theory is not true, and therefore the identity theory is false, which makes its simplicity moot. The foregoing discussion has briefly summarized issues that provide reasons to think the identity claim central to the identity theory is false. Next, I will specifically analyze the argument for the identity theory.

## Simplicity Argument Analysis

Having discussed reasons to think the identity theory is false, which at least provide motivation for considering alternative views, I will now critique Smart's simplicity argument for the identity theory. In this section, a critical analysis of the simplicity argument will be given, but two points of clarification are in order.

For one, the principle of simplicity and the simplicity argument are not the same thing. The argument utilizes (or is based on) the principle. The principle of simplicity is merely a premise in the simplicity argument, not the argument as a whole. Moreover, successful and unsuccessful arguments can have one or more true premises. This brings me to the second point: The epistemic principle of simplicity is indispensable. Not only is it theoretically well justified, its track record demonstrates that it is pragmatically useful. In this section, objections to the simplicity argument will be given, but these objections are not aimed at the principle of simplicity. The argument can fail to be cogent, even though one of its premises—the principle of simplicity—is true and indispensable.

As for my critical analysis of the argument, I'll highlight three concerns. First, I will discuss the fact that there are additional theoretical virtues to consider. My second concern pertains to the scope of data being explained. Lastly, I will discuss the simplicity argument's misrepresentation of dualism.

### Theoretical Virtues

One weakness of the simplicity argument is that it rests upon one, and only one, theoretical virtue. Simplicity is significant, but it's not the only theoretical virtue. To use Mario Bunge's (1961, p. 120) terminology from *The Weight of Simplicity in the Construction and Assaying of Scientific Theories*, simplicity "competes" with additional "desiderata." There are additional theoretical virtues to consider, such as explanatory power, explanatory scope, fertility, internal coherence, and consistency with widely accepted theories and background knowledge.[27]

If the identity theory is the simplest theory, that's a significant point in its favor. However, by itself simplicity is not conclusive. If simplicity itself were conclusive, then we should go with a view like solipsism and conclude that there is just one mind that exists, which is imagining NCC data along with everything else. It's hard to think of a simpler theory. But given that there is more to consider than simplicity, we need not commit ourselves to solipsism simply because it is simple.

Likewise, simplicity itself is not enough to establish that dualism is false and the identity theory is true. Unless, of course, all else is equal and the identity theory is on par with dualism on every other score. If that's the case, then simplicity can act as an epistemic "tiebreaker" tipping the balance in favor of the simplest view. In other words, the simplicity argument requires another premise affirming the condition mentioned in premise one, which reads: *All else being equal*, the simplest explanation of a data set is the best explanation. The needed premise is something like: (EQUAL) All else is at least equal or in favor of the identity theory. So we can give a second, more accurate, reconstruction of the simplicity argument:

(SIMPLICITY)   All else being equal, the simplest explanation of a data set is the best explanation.
(EQUAL)        All else is at least equal or in favor of the identity theory.
(NCC-DATA)     Neuroscience has discovered NCC.
(ID-SIMPLER)   Relative to dualism, the identity theory is the simplest explanation of NCC.

*Therefore:*
(ID-BEST-EX)   The identity theory is the best explanation of NCC.

While a justifiable appeal to simplicity requires (EQUAL), some might think it needs no defense and that it is safe to assume. However, (EQUAL) is not obviously true, but questionable given the apparent differences between consciousness and neural correlates discussed above, which dualist views are often consistent with. And as discussed in chapter 1, I think there are good arguments offered by leading philosophers for dualism. Granted, dualism has its problems (which is why I intend to demonstrate in this work that neo-Thomistic hylomorphism is capable of overcoming such problems). My point, however, is merely that (EQUAL) is questionable. It is also where much of the debate actually lies. Those inclined to either dualism or physicalism will often consider this premise in the light of differing metaphysical presuppositions that influence the apparent plausibility of (EQUAL).[28]

*Data Scope*

Another limitation of the simplicity argument is that it appeals to a narrow scope of data, as it focuses merely on the fact that there are NCC. As alluded to above, Christof Koch pioneered NCC research with Francis Crick. Koch (2019b, pp. 48–49) is very clear about the limitations and neutrality of NCC with respect to the metaphysics of mind:

> Francis Crick and I meant the NCC language to be *ontologically neutral* (which is why we spoke of correlates) with regard to the age old battles of the -isms (dualism versus physicalism and their many variants . . .), as we felt that at this point in time, science could not take a firm position with respect to resolving the mind-body problem. No matter what you believe about the mind, there is no doubt that it is intimately related to the brain. The NCC is about where and how this intimacy takes place.

The data revealing that there are NCC is well established and will become more so as the contemporary search for neural correlates continues. This information is beneficial for treating mental ailments with a neurobiological basis. However, with respect to the nature of the mind and consciousness, NCC data itself is limited and underdeterminative.

If an argument appeals to empirical data, you want the scope of the data to be as wide as possible. The simplicity argument makes such an appeal. A key premise is:

(NCC-DATA)    Neuroscience has discovered NCC.

The problem is that such information provides a very narrow scope of data when applied to debates about the nature of the mind and consciousness. Physicalists might be naturally inclined to think that a wider data scope gained from future neuroscience research will just further support physicalism. After all, they think their view accurately describes reality as it pertains to the mind, and a wider data scope will only tell us more about reality in that area. Although for the exact same reason, nonphysicalists might reasonably expect a wider data scope to confirm their view (see, e.g., LaRock et al. 2020). Again, one's presuppositions can play an epistemic role, in this case by influencing expectations of what a wider data set will support. In any event, it is unclear what future empirical data would support physicalism, and it is to the simplicity argument's detriment that it appeals to such a narrow scope of empirical data.

It is also worth pointing out that the type of data appealed to in the simplicity argument is limited to just neuroscientific data. We have more data about ourselves than just neuroscientific data (which we have only gained in recent times). We also have what Roderick Chisholm (1976, pp. 16–18) called "philosophical data" about ourselves. Such data includes our capacity to reason, to think about things, to make choices, our persistence through time and bodily changes, our moral awareness, and so on. If a theory provides the simplest explanation of the fact that there are neuronal correlates of our conscious experiences, that's a point in its favor. However, that does not say much about how well it can account for philosophical data about human persons. When such data is also considered, reductive physicalism's simplicity can begin to appear more like poverty (cf. Koch 2012, p. 152).

In sum, a considerable weakness of the simplicity argument is that it appeals to a narrow scope of data—merely the neuroscientific discoveries of NCC. To conclude that dualism is false and the identity theory is true based on such a narrow data set is simply too hasty, despite the identity theory's simplicity.

## Dualism Misrepresented

The version of dualism targeted by the proponent of the simplicity argument is often a minimal version of dualism, simply the idea that mental states are irreducible and nonphysical. And, in light of NCC, it is thought that laws relating these irreducible nonphysical mental states to neural events must be postulated. This sort of dualism seems to be the target of Smart's (1959, pp. 142–143) presentation of the simplicity argument as well as Hill's (1991, p. 20). It is easy to understand the motivation for targeting a minimal version of dualism. The idea is that if you can undermine the minimal tenet(s) common to all versions of dualism, then you undermine all versions of dualism.

However, there are several problems with this approach. Firstly, the minimal version of a position is not always the most defensible version. Sometimes a more robust version of a position is more defensible precisely because it's more complex, or more nuanced, or has more resources available to deal with problems and provide explanations. Secondly, the minimal version of dualism that is the target of the simplicity argument misrepresents basic elements of the version of dualism I will defend in later chapters. On the view I'll put forth, there are not countless individual isolated mental states that together compose the mind. Rather, there is one simple substance that engages in mental activity by exercising mental powers.[29]

Thirdly, proponents of the simplicity argument seem to think that dualism is a neuroscience theory meant to account for neuroscientific data, which is not the case. Most dualist philosophers are not dualists because they think dualism is the best theory for explaining NCC or any other neuroscientific data.[30] Typically, the rationale for dualism is strictly or principally philosophical and not empirical. For example, the nonphysical mind of substance dualism is thought to be what grounds human essence (cf. Oderberg 2005), personal identity (cf. Lowe 2001), subjective experience (cf. Zimmerman 2011), agency (cf. Swinburne 2013), intentional mental causation (cf. Lowe

2006), cognition (cf. Fumerton 2013), and/or moral awareness (cf. Swinburne 1986).

Since the nonphysical mind of substance dualism is not presented as a neuroscientific theory by dualists or argued for as if it were, it should not be evaluated principally on how good of a neuroscientific theory it is. With that said, however, in chapter 7 I'll try to show how neo-Thomistic hylomorphism can account for NCC. Yet my aim will be to defeat a potential defeater of dualism, not to give a positive case per se for dualism.

## CONCLUSION

The claim that neuroscience nullifies dualism is often thought to be supported by the discovery of neural correlates of consciousness. To the contrary, I have argued in this chapter that such correlations don't entail any particular view of the mind. Moving from the fact of NCC to the conclusion that physicalism is true and dualism is false is not an empirical step. Rather, it requires philosophical argumentation, which the simplicity argument aims to provide. However, there are reasons to think the conclusion of the simplicity argument—the type identity theory—is false. Furthermore, the argument for the conclusion has considerable weaknesses. Thus, contrary to popular opinion, it is not clear how NCC are supposed to invalidate dualism. Moreover, I'll argue in chapter 7 that my neo-Thomistic hylomorphic position, which is a version of dualism, provides a good explanation of NCC.

However, there is a different issue threatening dualism's viability—that is, mental causation. In fact, it is this issue that is cited by Koch (2012, p. 151) when he aims to falsify substance dualism in *Consciousness: Confessions of a Romantic Reductionist*:

> Two recent defenders of dualism, the philosopher Karl Popper and the neurophysiologist and Nobel laureate John Eccles, made an appearance in chapter 7. Let me repeat a point I made there when discussing their views on Libertarian free will. The dualism they advocate, in which the mind forces the brain to do its bidding, is unsatisfactory for the reason that the 25-year-old Princess Elisabeth of Bohemia had already pointed out to Descartes three centuries earlier—by what means does the immaterial soul direct the physical brain to accomplish its aim?

The problem regarding mental causation that Koch refers to has been developed by Jaegwon Kim. Kim (2005, p. 74) alleges that the argumentation of Descartes's earlier critics, including Princess Elisabeth, falls short even though it is on the right track. Therefore, he developed the point made by

Princess Elisabeth into what is known as the causal pairing problem (see Kim 2005, Ch. 3). This problem is dualism's most daunting difficulty regarding mental causation.

Therefore, addressing the causal pairing problem is one of my aims in this work. However, there are two other problems regarding mental causation that do not threaten dualism like the causal pairing problem does. So before addressing the causal pairing problem itself, I will clarify and discuss these two other problems in the next chapter for the sake of distinguishing dualism's Annapurna with respect to mental causation, which is the pairing problem.[31]

## NOTES

1. Cf. Mitchell Glickstein (2014, p. 1).

2. Horton seems to be disavowing dualism and advocating materialism. The previous line says: "Philosophical defenses of materialism seem increasingly substantiated by science" (Horton 2011, p. 376). Following that line, he says neuroscience has proven that the mind is matter. However, on the next page he advocates what he calls "dichotomy," according to which the soul is distinct from the body and persists after bodily death (Horton 2011, p. 377). This seems very dualistic. Moreover, on the same page he has a footnote where he commends John W. Cooper's (1989) position, which Cooper calls "holistic dualism" (see Cooper 1989, Ch. 10 section IV). And elsewhere Cooper (2009, p. 46) commends "Thomist dualism."

3. C-fiber activation correlated with pain is the standard example in the philosophical literature; therefore, I'll often use this example. However, as Christof Koch pointed out to me regarding C-fibers in personal correspondence: "Ironically, I and most other neuroscientists would *not* consider these a true content-specific NCC; just like the optic nerve isn't a visual NCC. Those are input structures that convey action potentials to cortex. The pain-NCC is higher-order somato-sensory cortex as phantom pain demonstrates" (email October 8, 2017; quotation used with permission).

4. For examples of research regarding the full NCC, see Maquet et al. (1997), Massimini et al. (2005), and Alkire et al. (2008). For examples of research regarding content-specific NCC, see Blake and Logothetis (2002), Tsuchiya and Koch (2005), and Lamme (2020).

5. Since there are multiple types of sufficiency and necessity, further inquiry should take this into account. For an in-depth analysis of how "sufficient" should be understood in the definition of NCC, see M. Owen and Guta (2019).

6. Koch et al. (2016, p. 308) give a similar definition: "The NCC are defined as the minimum neuronal mechanisms jointly sufficient for any one specific conscious percept." See also Koch (2019b, p. 48).

7. Chalmers (2000, p. 32) points out that lesion studies are often used to make inferences about NCC, but he thinks such methodology is flawed. According to

Chalmers, "The identity of an NCC is arguably always relative to specific brain architecture and normal brain functioning, and correlation across abnormal cases should not generally be expected." What was an NCC can cease to be such when a lesion alters brain structure, thus warranting caution when making inferences from lesion studies. Fully aware of the case (and need) for caution, Christof Koch pointed out to me in conversation (October 16, 2017) that lesion studies can nevertheless provide important information when it comes to identifying NCC especially when such information is coupled with findings from artificial stimulation studies in a healthy brain that corroborate lesion studies (see, e.g., Koch et al. 2016, p. 308).

8. I'm indebted to Koch for this point and the foregoing subpoints elucidating the overall idea. As he pointed out to me in conversation (September 11, 2017), there can be variations pertaining to NCC due to variations of a subject's overall conscious experience and differences in individual brains. Koch was then the president and chief scientific officer at the Allen Institute for Brain Science, which endeavors to map the human brain and mouse brain (see brain-map.org). (Koch is now the chief scientist of the Allen Institute's Mindscope Program.) The Allen Institute's research on the mouse brain does not generally involve cloned mice but highly inbred mice that are isogenic. Although their genomes are very similar, they're not quite the same, which is also true of cloned mice due to biological noise, or subtle inconsistencies.

9. I am here, and throughout the rest of this section, indebted to an anonymous reviewer for their very insightful suggestions on an earlier draft.

10. For brief introductions to binocular rivalry and the relevance to NCC, see Mormann and Koch (2007, section "The Neuronal Basis of Conscious Perception") as well as Blake and Tong (2008).

11. Here I've benefited from conversation (March 15, 2017) with Koch about the motivation for the no-report paradigm.

12. Block and Stalnaker (see 1999, pp. 23–25) make an important point regarding explanatory expectations the advocate of the identity theory might appeal to in response to various articulations of the third objection Smart (1959, p. 148, footnote 11) considered and described as "the one which I am least confident of having satisfactorily met." See the section "The Identity Theory's Viability" below.

13. However that's not to say, according to Smart, that "pain" means the same as "C-fiber activation" (see Smart 1959, pp. 144–145). Accordingly, one cannot infer that pain is identical to C-fiber activation from the meaning of terms; it's allegedly empirically discovered.

14. Another interpretation of Smart says that irreducible conscious states are "danglers" since they are epiphenomenal (cf. Feigl 1967; Polgar 2011; Smart 2007). This makes Smart's argument hinge on the idea that irreducible mental states are necessarily epiphenomenal. Many dualists disagree, so Smart would need to show that such states are necessarily epiphenomenal. Some think causal closure entails such. In chapter 3, I address closure. Kim (2011, Ch. 4) has argued that because nonphysical mental states would be epiphenomenal, we should reduce them to neural correlates. Kim relies on the causal pairing problem to support the premise that nonphysical states would be epiphenomenal (see Kim 2011, p. 113, endnote 15). I present the pairing problem in chapter 4 and respond to it in chapter 6.

15. Cf. Nagel (2012, pp. 39–40).

16. For further sources pertaining to issues discussed in summary fashion here, see Chalmers (1996), Jackson (1982; 1986), Kripke (1981), Block and Stalnaker (1999), Robinson (2012), and Swinburne (2013).

17. For an introductory summary of Leibniz's law, see Beebee et al. (2011, pp. 114–115).

18. Multiple realizability raises a problem for Smart's type identity theory, but not token identity theories.

19. IIT claims that the physical substrate of consciousness is an integrated structure in the central nervous system that exhibits maximal intrinsic cause-effect power (Tononi et al. 2016, p. 450). The theory is discussed further in chapters 7 and 8.

20. For notable replies to the line of reasoning presented here, see Churchland (1985), Tye (1986), and Loar (1990).

21. This is true even on the "no-report" paradigm. Reports from the subjects are still used to identify physiological indicators of consciousness and in the trials that are compared to the study trials that rely on the physiological indicators (see the section "What is an NCC?").

22. One might very well disagree with IIT's axioms (see, e.g., Bayne 2018). But the point is simply about the starting point of beginning with what we know directly about our conscious experience rather than what we know indirectly through empirical investigation about the brain.

23. For worries about zombies, such as whether they're conceivable and what that would entail, see Dennett (1995), Marcus (2004), and Bailey (2009).

24. However, a recent study by Aparna Bhaduri et al. (2020) raises questions about the extent to which cerebral organoids actually resemble the human brain (see Weiler 2020).

25. See Lavazza (2021) and Reardon (2020).

26. See Hacker (2007, p. 252) for a concise list that includes other apparent differences and related issues.

27. While most of these theoretical virtues are commonplace, fertility is less known. If a theory leads to further research opportunities, it has the theoretical virtue of fertility.

28. Cf. Tahko (2012, pp. 40–41).

29. This view, presented and defended in chapters 5–8, is also arguably much simpler than a physicalist account of human persons that says we and our mental lives are constituted out of countless fundamental physical substances (whether they be atoms or subatomic particles).

30. I am indebted here to J. P. Moreland, who has often made this point, which is a general point that permits exceptions.

31. A forerunner of this chapter was published in *Consciousness, and the Ontology of Properties*, edited by Mihretu Guta and published by Routledge. Copyright © 2018 from *Consciousness, and the Ontology of Properties*, edited by Mihretu Guta. Reproduced by permission of Taylor and Francis Group, LLC, a division of Informa PLC. This permission does not cover any third-party copyrighted work which may appear in the material requested (user is responsible for obtaining permission for such material separately from this grant).

*Chapter 3*

# Mental Causation
## *Identifying Dualism's Problem*

It's the bell lap of a one-mile footrace. Four hundred meters to go, and Roger Bannister's legs become heavy, as lactic acid is released into his blood stream as his muscles begin lacking oxygen. His body's physiology screams at him to slow his pace. Yet his mental desire to cross the finish line in less than four minutes causes the opposite effect—his stride only quickens. That's mental causation. Another example would be a father's intention to express love to his child being causally relevant to his arms opening wide. Mental causation in terms of top-down, mental to bodily causation, is the focus of most contemporary discussions about mental causation. The same is true of this work. Nevertheless, it is worth noting that there is bottom-up causation as well where mental events are caused by bodily events. For example, a pin pricking one's skin causes an experience of pain.

Mental causation, it seems, is a prerequisite for essential human capacities like rationality and agency. Nevertheless, accounting for mental causation is difficult for any view of the mind. The various problems are not merely puzzling but also consequential. Jerry Fodor (1989, p. 77) offers sobering words:

> If it isn't literally true that my wanting is causally responsible for my reaching, and my itching is causally responsible for my scratching, and my believing is causally responsible for saying . . . if none of that is literally true, then practically everything I believe about anything is false and it's the end of the world.

Given the consequential entailments, if mental causation is incoherent on a particular view, that view ought to be rejected. Many charge substance dualism with such incoherence, which its problem with mental causation allegedly demonstrates.

But what exactly is substance dualism's problem with respect to mental causation? A fact too often overlooked is that there are multiple problems pertaining to mental causation. Different problems arise given different assumptions. As Kim (2000, p. 29) points out:

> Philosophical problems do not arise in a vacuum. Typically they emerge when we come to see a conflict among the assumptions and presumptions that we explicitly or tacitly accept, or commitments that command our presumptive respect.

These words are found in a chapter entitled "The Many Problems of Mental Causation," which summarizes various causal problems entailed by various philosophical tenets. Different problems regarding mental causation arise from different tenets entailed by different views. As we will see, the causal exclusion problem arises due to what's called the "causal closure principle." As will also become clear, the causal pairing problem arises for Cartesian substance dualism since it denies causal closure and posits causal interaction between a nonphysical mind and a physical body.

While there are multiple problems regarding mental causation, my aim is not to deal with every one of them. Rather, I will ultimately deal with one problem that is substance dualism's chief problem regarding mental causation—the causal pairing problem. The pairing problem charges substance dualism with an incoherence pertaining to mental causation. The purpose of this chapter is to clarify two other problems of interest to contemporary philosophers in order to distinguish them before subsequently setting them aside. The goal is to provide clarity and prevent confusion, by parsing out the two other problems that will not be my focus in subsequent chapters when I address specifically the causal pairing problem. However, avoiding real issues is not the goal. Therefore, justification will be given for why the causal exclusion problem and the problem of a lack of psychophysical laws will be set aside when the pairing problem later becomes my focus.

In the section "Lack of Psychophysical Laws," the problem of a lack of psychophysical laws will be explained and my rationale for setting it aside will be given. In the section "Causal Exclusion Problem," the causal exclusion problem will be presented, and I will discuss in some depth why the substance dualist's denial of the causal closure principle is warranted. My rationale will be aimed at questioning the physicalist's justification for thinking that closure is demonstrably true, yet I will also offer reasons to think that the principle is false. A denial of causal closure, if warranted, circumvents any threat the exclusion problem poses for substance dualism. Ultimately, this chapter is intended to elucidate why I focus on the pairing problem in later chapters. With that said, let's consider the first of these *other* two problems.

## LACK OF PSYCHOPHYSICAL LAWS

The problem of lacking psychophysical laws is connected to Donald Davidson's view of the mental, which is known as *anomalous monism*. In this section, I will analyze this view and the problem it prompts. To start with, anomalous monism and the way in which it leads to the problem of a lack of psychophysical laws will be presented. Then the set of assumptions that entail the problem will be specified. Lastly, rationale will be given for subsequently setting this problem aside.

On Davidson's view, mental events are events describable in mental terms and physical events are events describable in physical terms (see Davidson 2001, p. 215; cf. Glüer 2011, p. 250). According to his anomalous monism, mental event types are distinct from physical event types, and there are strict causal laws pertaining to physical types, but no such laws regarding mental types (see Glüer 2011, p. 252). Hence, Davidson thought that every token mental event that causes a particular physical event is identical to a token physical event (Glüer 2011, p. 250). Yet mental types, according to Davidson, are irreducible to physical types. For example, Emma's mental intention that causes her left arm to rise is identical to some instance of a physical event. However, the type of mental event, which Emma's intention is an instance of, is irreducible to any physical type. In other words, the mental type "intention to raise left arm" is not reducible to a specific physical type, such as "brain fiber-$\phi$ firing."

Why adopt anomalous monism? At the heart of Davidson's rationale is the idea that there are strict laws governing physical causation, which make causation possible between physical events, but there are no such laws governing mental to physical causation (Glüer 2011, pp. 248–256). In other words, there are no psychophysical laws—that is, laws that govern causation between mental events and physical events. This is problematic given two assumptions: (1) the causal framework of event causation, and (2) the nomological requirement. Those who hold to event causation think of causation as a relation between (or among) events.[1] Causation, on this framework, is an external relation requiring distinct events as relata that stand in relation to one another. The nomological assumption says that whenever one event causes another event, the causal relationship is derived merely from noncausal features of the situation and pertinent covering laws.[2] Given these assumptions, each case of causation has distinct events governed by a covering law(s).

Now we can understand the motivation for anomalous monism. Since there are no psychophysical laws, there cannot be causation between nonphysical mental events and physical events. Consequently, whenever a mental event causes a physical event, the mental event must be a physical event; otherwise, it could not cause the physical event. Reducing the specific mental event to a physical event allegedly allows one to explain why the mental event caused the physical event. Yet the idea is that mental types are still distinct from physical types. Supposedly, this view allows one to have mental causation without a full-blown reduction of the mental to the physical.

Given the above assumptions, and Davidson's belief that there can be no psychophysical laws, we can see why he adopted anomalous monism. But why did he think there cannot be psychophysical laws? Before we answer this question, let's be clear about what Davidson considered a "strict law" to be. According to Davidson (2001, p. 215), laws are linguistic. Strict laws, it's thought, are true statements that are universally applicable without qualification (Glüer 2011, p. 252). Such laws allow us to explain why events take place. Given the law of gravity, for example, we can say why a coin fell to the floor when I released it from my hand. Whenever anyone releases a coin, while on earth, the coin will fall toward the ground. The thought is that physical laws apply universally without exception, so we can explain physical events by appeal to such laws.

Not so for mental events (Davidson 2001b, p. 216). Davidson (2001b, p. 216) thought that mental concepts are "irreducibly causal," and thus cannot be specified in a way that universally applies (Glüer 2011, p. 254). This flows out of Davidson's position known as the "holism of the mental." According to which, what leads someone to act in a specific way has to do with their whole psychology at that time, not particular mental events. So psychological causes of actions are not specific mental states or specific reasons. Yet, Davidson also thought psychological explanations, which are needed for "strict laws," are aimed at specifying the specific reasons that cause someone to act.

Therefore, a problem arises: the specific reasons that cause one to perform a given action need to be specified for there to be strict laws, but such cannot be universally specified (Davidson 2001b, p. 216; Glüer 2011, p. 255). For one, we can never identify particular reasons that cause one to act because it is one's entire psychology at a time that causes their action, not specific mental states. In addition, the same reasons will not always cause the same result. Basically, strict laws depend on psychological explanations that require the identification of specific reasons that universally cause specific actions, but such identification is not possible given the holism of the mental.

To make matters clearer, let us consider an example. On April 19, 1861, one week after Confederate forces fired at Fort Sumter in Charleston Harbor, President Abraham Lincoln ordered a blockade of southern seaports. In essence, Lincoln's order began the United States Civil War. What reasons spinning around in Lincoln's mind caused his act? Consider these hypothetical reasons: (*a*) the blockade will severally damage the economy of the rebelling states; (*b*) the blockade will prevent Confederate forces from securing needed military supplies; (*c*) instigating war with the South will lead to freedom for southern slaves; (*d*) being president during a civil war will secure Lincoln's place in history; (*e*) not responding to the shots fired on Fort Sumter will send the message that the North is ill prepared for war.

Given the holism of the mental, it is impossible to specify which reason (or set of reasons) caused Lincoln to act as he did. For Lincoln's action was determined by Lincoln's entire mental psychology at the time in question, not by any one particular reason or set of reasons. Therefore, we cannot specify which reason(s) caused Lincoln to act as he did. Even if we could, there's an additional problem. The reasons that caused Lincoln's action would not cause the same effect without exception.

Suppose Lincoln's action was caused by reasons *a* and *c*. In such a case, for there to be a psychological law, we need to be able to say that *a* and *c* will always cause the same choice by someone in Lincoln's position. Although if Lincoln had one additional belief—for example: (*f*) a blockade of southern ports will also damage the economy of potential allies—this may have altered his decision entirely. Additionally, suppose the First Lady, Mary Lincoln, was President and Abe was the first, First Gentleman. We cannot say that she would have done the same, and we especially cannot say whether reasons *a* and *c* would have caused her to do so. Psychological explanations are not like physical explanations. We cannot specify which reasons will universally, and without exception, lead to certain actions. Therefore, according to Davidson's line of reasoning, there are no strict psychological laws (Davidson 2001b, p. 216; Glüer 2011, pp. 255–256). For such laws would require us to be able to say which reasons will cause specific actions universally, without exception. We cannot do so. Therefore, according Davidson, there are no psychophysical laws that causally relate mental events to physical events.

That is the problem of a lack of psychophysical laws and the rationale for it. Given that there are no psychophysical laws, the alleged problem for dualism is that there are no laws to relate nonphysical mental causes to physical effects. Such laws are supposedly necessary for one event to cause another event. Therefore, the thought is, there cannot be causation between nonphysical mental events and physical events.

As mentioned above, this work is not focused on responding to this problem. Rather this problem has been summarized and clarified for the sake of preventing confusion in later chapters. This problem—the lack of psychophysical laws—will be subsequently set aside and following chapters will not address it. Nevertheless, before we set the problem aside, let's consider warrant for doing so.

## Justification for Setting the Problem Aside

I will not be focusing on the problem of a lack of psychophysical laws in the following chapters because I intend to deal with the chief objection to substance dualism based on causal grounds. This problem does not present such an objection. After all, philosophical problems arise given certain tenets. I have outlined tenets Davidson held that led him to conclude that there are no psychophysical laws. However, dualists are not necessarily committed to the same tenets or assumptions that produce the problem. Dualism does not commit one to the following assumptions held by Davidson:

a. Causation is a relation between events.
b. Causation is an external relation between causal relata (i.e., events).
c. Causation requires strict universal laws to explain cause and effect.
d. Laws are linguistic and depend on our identification of universal regularities.
e. Reasons one has for a particular belief or action cause one's belief or action.

Minus any one of these assumptions, the problem does not arise. Moreover, dualists do not need to accept any of these assumptions, unless they are given convincing rationale.

Many dualists are disinclined to accept the framework of event causation that says causation is a relation between events, and rather endorse agent causation (cf. Lowe 2013; Plantinga 1984). On the framework of agent causation, agents have causal power, and effects do not necessarily depend on previous events. One who accepts agent causation can endorse the notion of pure powers, which don't rely on other causal powers to make them powerful (see Marmodoro 2010). Given pure powers, one would not need to think causation is necessarily an external relation between cause and effect. Consequently, the need for universal laws connecting cause and effect would be moot. Dualists are not tied down to the idea that there must be strict universal laws to account for mental causation (see John Foster 1991). We could go on listing points of disagreement dualists have with Davidson's assumptions. But I will mention just one more. The dualists need not think that one's reasons for acting cause their action. Thus, it wouldn't necessarily matter if there are no laws causally relating particular reasons to particular actions. In short, Davidson's problem arises due to the tenets he holds and very few, if any, of his tenets are necessitated by dualism. Therefore, the problem of a lack of psychophysical laws does not necessarily arise for the dualist.

However, even if dualism's tenets did give rise to the problem, it is unclear how it would pose a serious threat. According to Davidson's line of thought, strict laws are essentially statements with universal applicability, which depend on our ability to ascertain why, how, and when certain events cause other events. Laws basically boil down to our linguistic statements that fittingly apply. If that's what laws are, the lack of them is inconsequential. It could be that whatever makes possible causation between a mental state and a bodily state is unascertainable. Yet, nothing follows that compromises mental causation on dualism. We may never know why, or how, pinpricks cause pain events. Likewise, we may never know why, or how, my intention to raise my arm results in my arm rising. Nevertheless, not knowing the answers to the "why question" or the "how question" would not threaten the fact that there is something that makes possible the consistency of pinpricks causing pain events and intentions causing bodily movements. Facts about the world can be entirely unknown to us or inexpressible for us. Our inability to specify why, how, or when certain mental states will lead to physical states does nothing to undermine the fact that they do. Therefore, a lack of psychophysical laws, in Davidson's sense of "law," seems harmless to dualism.

To summarize, in this section we have laid out the problem of a lack of psychophysical laws and the rationale behind it. Furthermore, two important points have been made in an effort to warrant setting this problem aside. The first point is that the problem of lacking psychophysical laws rests on a number of tenets dualists are disinclined to accept. Admittedly, for the one who finds Davidson's tenets compelling, giving up any of them will entail some cost. Nevertheless, the multiple ways of avoiding the problem merely require a denial of one of the tenets that the problem depends on. Substance dualists can deny any of Davidson's assumptions and be perfectly consistent with dualism. The second point is that the problem of a lack of psychophysical laws in Davidson's sense of "law" does not offer a clear threat to dualism. In light of these points, I will set this problem aside in an effort to deal in depth with dualism's chief causal problem (the causal pairing problem).

However, before doing so there is another problem to be clarified and, again, subsequently set aside. The justification for doing so, however, will demand more extensive considerations. We now turn our attention to one of the most notorious problems regarding mental causation.

## CAUSAL EXCLUSION PROBLEM

The causal exclusion problem boils down to the mental being excluded from playing a genuine causal role. Given that every physical event has a sufficient physical cause, mental causes are excluded from causing physical events. Loosely speaking: all causal jobs are filled by physical causes, so mental causes are out of work. Technically speaking: if physical event $p$ is sufficiently caused by mental event $M$, the causal closure of the physical domain will be violated; yet if $p$ has a sufficient physical cause, say $P^*$, then $P^*$ would preempt $M$ as the cause of $p$. In effect, $P^*$ excludes $M$ as the cause of $p$.[3]

That is the causal exclusion problem. In this section, the causal exclusion problem and what gives rise to it will be explicated. Additionally, I will consider two routes that can be taken to avoid the problem, one of which I'll endorse. The endorsed route hinges on a controversial denial of what is known as the *causal closure principle*. Hence, considerable space will be devoted to rationale for such a denial.

Let us begin by considering an example. Timothy's mother is teaching him to give charitably. Before church, where they are invited to give charitably, Timothy's mother hooks him up to an electrical system that shocks him. The electrical shock causes his hand to open and release money held in it. When the offering plate passes by young Timothy, his mother hits a button that sends electrical currents through his body causing his hand to open, and out falls the money into the plate. Given the way things are set up, the money falls into the plate due entirely to the physical causes.

Timothy's mother needs parenting classes, but she might also benefit from a critical thinking class. After all, she is trying to teach her son to be charitable through a causal system that leaves no room for genuine charity to play a causal role. The physical causes—the electrical current and the automatic neural and muscular reactions—sufficiently explain why the money falls into the offering plate. There's no room for a charitable attitude, desire, or intention of Timothy's to play a meaningful causal role in bringing about the effect of the money falling into the plate. For even if Timothy desires to give to the church that feeds the poor, his desire is not necessary to cause his hand to open and drop the money into the plate. The money will fall into the plate wholly apart from Timothy's charitable desire. Whether his desire is present or not, the physical causes bring about the effect. Here's the problem: Timothy's mental states are excluded from having a causal impact.

Given four principles, the same problem arises for mental causation in general. Scott Sturgeon (1998, pp. 413–414) defines these principles as follows:

- (COP) *Completeness-of-Physics:* Every physical effect has a fully revealing, purely physical history.
- (IMP) *Impact-of-the-Mental:* Mental events have physical effects.
- (NOD) *No-Overdetermination:* The physical effects of mental events are not generally overdetermined.
- (DUAL) *Dualism:* Mental events are not physical events.

(COP) is also known as the causal closure of the physical domain. According to it, every physical effect has a sufficient physical cause. It is often thought that this principle is supported by the physical sciences. (IMP) essentially says that the mental aspects of human persons cause physical effects. It is commonsensical that my mental belief "a rock is about to fall on me" plays a causal role in bringing about physical effects such as my legs moving in a running motion. According to (NOD), physical effects that are mentally caused are not caused by more than one sufficient cause. In most cases, one sufficient cause does the job. (DUAL) is the idea that mental events are not identical to physical events. This principle is consistent with a nonreductive physicalism that says the mental cannot be reduced to the physical, a property dualism that says mental properties are sui generis properties, and substance dualism which claims human persons have or are nonphysical minds.

Any set of three of these principles is consistent. Yet the set of four is inconsistent (see Sturgeon 1998, p. 414). Here's why. If it were true that every physical effect had a fully revealing purely physical causal history, then no mental cause could play a causal role in bringing about any physical effect as long as mental events are distinct from physical events. One might think that physical events could just have multiple sufficient causes, like a ship that sinks due to multiple holes that could sink the ship on their own. But if this were so, there would be overdetermination, and (NOD) would be false. It seems the set of all four principles is simply inconsistent.

Yet there are four simple ways to avoid this inconsistency. Since any set of three of the principles is consistent, a denial of any one of the four allows one to dodge the incoherence. The epiphenomenalist is content with the idea that the mental is causally inefficacious, and thus denies (IMP). The physicalist has no qualms with a purely physical world with merely physical people, and therefore denies (DUAL). Dualists may have options. At first glance, it appears that they can avoid the inconsistency by choosing one of two routes:

> *Route OD*: posit that physical effects are often overdetermined by sufficient mental causes and sufficient physical causes; thus denying (NOD).
>
> *Route CO*: admit causal openness between the mental and the physical that allows nonphysical causes to have physical effects; thus denying (COP).

There are different motivations for taking either route. Let's consider each route in turn.

## *Route OD: Positing Overdetermination*

*Route OD* seems to be preferable for nonreductive physicalists who think the mental can't be reduced to the physical, and thus endorse (DUAL). For the alternative route, *Route CO* explicitly requires a denial of causal closure, which many physicalists believe is essential to a physicalist worldview (see Kim 1993, pp. 209–210). So *Route OD* is attractive to physicalists insofar as it doesn't ask one to deny anything essential to orthodox physicalism. However, whether or not positing overdetermination, and thus taking *Route*

OD, is an available option for the nonreductive physicalist will depend on how she views the closure principle. According to Kim (2000, p. 40):

> One way of stating the principle of physical causal closure is this: If you pick any physical event and trace out it's causal ancestry or posterity, that will never take you outside the physical domain. That is, no causal chain will ever cross the boundary between the physical and the nonphysical.

This articulation of closure precludes overdetermination. For if no causal chain ever crosses the boundary between the two domains, then physical effects simply could not have nonphysical causes that overdetermine them.

As a result, if the nonreductive physicalist views closure as Kim does, then the nonreductive physicalist seems to be in a bind. Neither *Route CO* nor *Route OD* can be taken without sacrificing closure, and so she seems stuck with denying either (DUAL) or (IMP). Given this, Kim (2000, p. 30) appropriately acknowledges the weight of this problem for physicalists.

> This, I claim, is our principal problem of mental causation. In referring to this as "our" problem of mental causation, what I mean to suggest is that it is a problem that arises for anyone with the kind of broadly physicalist outlook that many philosophers, including myself, find compelling or, at least, plausible and attractive. . . . The exclusion problem is distinctive in that it strikes at the very heart of physicalism.

Kim thinks physicalists can take one of two ways out: reductionism or epiphenomenalism (see Kim 2005, pp. 70–71). Other physicalists disagree with Kim and believe that there are versions of causal closure that are consistent with overdetermination.

Since my aim is not to defend any broadly physicalist view, but rather substance dualism, I will leave that debate for physicalists to hash out. With respect to substance dualism, I'm inclined to think that overdetermination will lead to problems when it comes to certain types of actions performed by agents where mental causes are inherently necessary causes, not just overdetermining sufficient causes. Therefore, it seems to me, *Route CO* is worth serious consideration.

### Route CO: Denying Causal Closure

*Route CO* requires a denial of causal closure—that is, a denial of a tenet that is in no way entailed by substance dualism. It's important to note that the causal pairing problem (which will be presented in chapter 4) would not arise for dualists if they did not deny causal closure. In other words, if dualists did not mix "physical and nonphysical events in a single causal chain," to borrow Kim's

(2000, p. 37) wording, there would be no need to pair nonphysical causes with physical effects. But this need provides the basis for the pairing problem.

Given that causal closure is in no way essential to substance dualism, it's no surprise that dualists often explicitly, or tacitly, deny the principle. Examples include Richard Swinburne (2013), Alvin Plantinga (2008), Robert Garcia (2014), Karl Popper and John Eccles (1983), George F. R. Ellis (2015, p. 27; 2016, pp. 13–14), and John Foster (1991). E.J. Lowe (2006, p. 18) has presented a dualist account of mental causation that is allegedly consistent with causal closure; nevertheless, even he thought the principle is false (see Lowe 2003, p. 145). After arguing that it's false, Lowe (2013, p. 169) pointed out that it need not concern him that his dualistic account of human agency is inconsistent with it. After all, closure is not necessitated by dualism.

Since the substance dualist can deny the causal closure of the physical domain without giving up anything essential to her view, *Route CO* is appealing. Yet we must consider whether or not dualists have sufficient warrant for denying closure. In the following section, we will consider the dualist's rationale for denying the causal closure principle.

## Rationally Denying Closure

Before moving forward, let's take stock of where we are. So far, we have considered the alleged problem of a lack of psychophysical laws. For various reasons, I set it aside. We then moved on and presented the causal exclusion problem. Two routes that one could take in order to avoid the problem while maintaining the idea that mental events are not physical events presented themselves. *Route OD* requires positing overdetermination. *Route CO* requires a denial of the causal closure of the physical domain. While this principle may be essential to physicalism, it is not part of substance dualism. Dualists can take *Route CO* without sacrificing anything essential to their view. Now we will turn to consider whether or not the dualist can rationally deny closure, and I will argue that the dualist is warranted in denying the causal closure principle.

Consider the following words from Kim (2003, p. 65):

> We commonly think we, as persons, have both a mental and a bodily dimension—or, if you prefer, mental aspects and material aspects. Something like this dualism of personhood, I believe, is common lore shared across most cultures and religious traditions.

Kim, a longtime proponent of physicalism, goes on to argue that this common dualistic conception of ourselves is false. Nevertheless, his claim that a dualistic conception of human ontology is commonly held seems correct (cf. Moreland

2011, pp. 21–22). After all, such is necessary for the possibility of disembodied existence after the death of our bodies or reincarnation. If we are identical with our bodies, both are obviously impossible. Yet countless people hold such beliefs. Hence, it is fitting that many people believe we have a nonphysical mental aspect, that is, a soul. Moreover, it seems to us that the nonphysical mental aspect brings about physical bodily effects. This is why epiphenomenalism "strikes most of us as obviously wrong, if not incoherent; the idea that our thoughts, wants, and intentions might lack causal efficacy of any kind is deeply troubling, going against everything we believe about ourselves as agents and cognizers" (Kim 2005, p. 70). In short, it seems to most people that they have a nonphysical mental aspect and that the mental has causal power.

As Roderick Chisholm (1976, p. 16) once pointed out, that which we are justified in assuming when not philosophizing, we are also justified in assuming when doing philosophy. Of course, such assumptions are not infallible. They can be proven false. The commonly held dualistic conception of human persons is a fair place to start, but it can be falsified by sound arguments. The causal exclusion argument allegedly does just that.

At this point, for the sake of clarity, let's formalize the causal exclusion argument against (DUAL) and for reductive physicalism in the following way[4]:

(IMP) Mental events have physical effects.
(NOD) Physical effects of mental events are not generally overdetermined.
(COP) Every physical effect has a fully revealing, purely physical history.
∴ ¬(DUAL) Mental events must be identical to physical events.

The physicalist relies on (COP) to secure the conclusion that (DUAL) is false and reductive physicalism is true.

Clearly, if physicalism is true, (COP) makes sense. Thus, the argument is forceful for those who accept a physicalist view. But according to the commonly held dualist view, (COP) seems false. For it seems that our nonphysical mental aspect causes physical effects all the time. The rock climber's trust that her rope will hold her and her intention to rappel down the cliff causes her grip of the ledge to release. The expectation that a specific person far away will receive a signal from their phone when a particular number is dialed leads to one's fingers pushing specific buttons. The skydiver's desire for a thrill and confidence that her parachute will open causes her to physically jump out of a plane. The judge believes that the terrorist bomber consciously intended to cause an explosion in order to kill innocent citizens, so she sentences the terrorist to life in prison while ignoring the flames causal role in igniting the dynamite, as if a mental intention was needed to cause a specific physical effect.

In short, it appears plausible prima facie that we have a nonphysical mind that often causes physical effects, and therefore causal closure seems false. Moreover, most philosophers that hold to substance dualism think there are further reasons that suggest we have a nonphysical mental aspect with causal power to produce bodily effects (see Plantinga 1984; 2012; Swinburne 2013). Rationale for this view would serve as justification for the falsity of the closure principle. What is important to note is that the substance dualist comes to the exclusion problem with a very different outlook than the physicalist. Given the physicalist's worldview, closure may seem obviously true. But given the substance dualist's worldview, which is not without warrant (see chapter 1, section "Presuppositions"), closure seems obviously false. For on dualism, the world is chalk-full of nonphysical minds that bring about physical effects all the time.

We must ask: What obliges the substance dualist to accept causal closure? The dualist is obliged to accept it if, and only if, the physicalist can *demonstrate* that closure is *true*. To do so, the physicalist must give compelling reasons that do not presuppose physicalism. Let's consider how the physicalist might show that the causal closure principle is demonstrably true.

## Analyzing the Justification for Closure

In this section, I will critically analyze justification physicalists commonly give for the causal closure principle. Throughout this section, I aim to give four primary points of rebuttal to the physicalist's rationale for closure. The first point will be that it's not easy to say what the principle actually claims, and, therefore, it's difficult to say whether or not it is justified by support offered in its favor. The second point is that empirical observation falls short of demonstrating that closure is true. The third point is that the success of science does not necessitate the truth of the causal closure principle. The fourth and final point will be that there are positive reasons to think closure is false. Each point will be considered in turn.

### *Ambiguity of the Causal Closure Principle*

A prerequisite for justifying a claim is specifying the claim to be justified. If you don't know what the claim is, then you can't know whether rationale given for it actually supports it or not. Thus, the first step in justifying the causal closure principle is clarifying what exactly it says. That is no easy task. As we shall see, various formulations of the principle and the variety of ways a key term in the principle can be understood make satisfying this prerequisite quite difficult.

After surveying five different versions of the principle, E. J. Lowe (2000, p. 574) aptly remarked: "One might have hoped for more exactitude and agreement amongst physicalists when it comes to the formulation of a principle so central to their position." Unfortunately, there is not one clear principle everyone has in mind in the conversation about causal closure. We have already noticed above that Sturgeon's definition of closure differs from Kim's and that the former is consistent with overdetermination whereas the latter is not. Here's a small sampling of further versions of the principle:

*Every physical effect has a sufficient physical cause.* Papineau (1998, p. 375) has presented this version. In contrast to Kim's (2000, p. 40) articulation quoted above, this version is compatible with overdetermination (Garcia 2014, p. 99).

*The chances of physical effects are always fixed by sufficient physical causes.* In an endnote following the above version, Papineau (1998, p. 386) assumes his reader asks about quantum indeterminacy and he indicates that closure can be put this way. More so than the above version, this version suggests determinism.

*At every time at which a physical event has a cause it has a sufficient physical cause.* Sophie Gibb (2013, p. 2) has put closure in these terms. This version also permits overdetermination.

*Every physical effect has its chance fully determined by physical events alone.* Martin Noordhof (1999, p. 367) prefers this version, which does not permit overdetermination and suggests determinism.

The various versions of the principle confuse matters. Although perhaps we could just pick one version that's agreed on by most physicalists in order to satisfy the prerequisite of specifying the principle to be justified. Kim (2009, p. 38) thinks most ontological physicalists will accept the following:

(COP*) If a physical event has a cause at $t$, it has a sufficient physical cause at $t$.

This principle seems simple and straightforward enough. Until we ask: What is meant by "physical"?[5] The answer is anything but straightforward. For consider the following puzzle.

Suppose that at a particular time $t_0$ you have two physical items. The first is a log with a circumference of 1 meter and height of 1 meter. The second is a flat square board with a length of 1 meter and a width of 1 meter. While it's clear that you have two physical objects at $t_0$, suppose that at a later time $t_1$ you place the log under the center of the board. So at $t_1$ the board rests on the log and forms a table (T). Do you now have three physical items—the log, the board, and a table? Or do you just have one physical item, the table? Or do you still just have the two original physical items, the log and the board? Various philosophers, even various physicalists, will give various answers. The same would be true if our example pertained to fundamental physical particles, rather than a log and a board. Such puzzles may seem irrelevant, yet there are genuine implications for the causal closure principle.

Sturgeon (1998) has pointed out that the term "physical" in the causal closure principle can mean *microphysical* or *macrophysical*. If microphysical is meant, then the claim is that every microphysical event has a sufficient microphysical cause. And what would count as "physical" would be fundamental physical particles (whatever they happen to be). If macrophysical is meant, the claim is that macrophysical events have sufficient macrophysical causes. And entities composed out of more basic fundamental physical particles could be concrete particulars that could causally produce effects. Simply put, if "physical" means microphysical, then our table (T) from our above puzzle doesn't count as a physical entity that exists, but if "physical" means macrophysical, then it does.

Therefore, if what counts as physical, according to (COP*), is microphysical fundamental physical particles, this appears to be inconsistent with the existence of macrophysical things such as planets, jet airplanes, animal bodies, biological organs, plants, automobiles, tornadoes, weather patterns, and so forth. These seem to be physical things, although they are not microphysical things. Thus, if the only things that count as physical entities are microphysical entities, according to the principle, then the principle seems to be at odds with the apparent existence of nonmicrophysical, macrophysical entities.

Furthermore, it is not just that macrophysical things seem to exist, but they also seem to causally produce physical effects that the principle does not address, if "physical" only means microphysical. Consider automobiles that collide into one another causing shattered glass and dented doors, jet engines that propel an airplane from one location to the next, a heart that pumps blood, a river that causes erosion, the movement of a runner's legs propelling her forward, and the like. It appears that these are physical entities, but not microphysical entities, which cause physical effects in the world. If "physical" in (COP*) only means microphysical, then the principle doesn't appear to address macrophysical causes of physical effects, which appear to exist. Moreover, it is also reasonable to think there are macrophysical causes of microphysical effects, which would be inconsistent with (COP*) if "physical" means microphysical. While there are many examples one might give, I'll give just two brief examples here of apparent macrophysical causes of microphysical effects that present prima facie difficulties.

First, consider a fundamental physical particle, $P$, which is sitting on the surface of my floor (let's call its location $L^1$) at time $t_1$. Suppose that I move a vacuum cleaner hose over $L^1$ and the vacuum inside the hose lifts $P$ upward into the hose causing it to change locations to $L^2$ inside the hose, where it is located at time $t_2$. It seems that the cause of $P$ moving from $L^1$ to $L^2$ is the vacuum inside the vacuum hose. It is difficult to reduce a vacuum to a fundamental physical entity because a vacuum requires a whole system of physical parts to be cooperative. The vacuum might be reducible to a structured collection of fundamental physical entities functioning in tandem. But such a collection is not a fundamental microphysical entity. Therefore, if it is the cause of $P$ moving from $L^1$ to $L^2$, it provides a counterexample to (COP*) if "physical" means microphysical.

Second, the natural function of a human brain requires countless individual oxygen molecules consisting of two oxygen atoms. For simplicity's sake, we can assume an atom is a fundamental physical entity, and thus microphysical. Through a physiological process—the repetition of systolic contractions—a human heart pumps blood that carries erythrocytes including hemoglobin with oxygen molecules consisting of oxygen atoms (see Martini et al. 2006, p. 68). We can imagine an individual oxygen atom, which is a constituent of an oxygen molecule, that is in a heart chamber at one time and then subsequently moves to the brain because the heart pumps the blood carrying the oxygen molecule including the individual atom to the brain. Without the heart (or an artificial heart) pumping the blood, the movement of such atoms to the brain would cease and the natural function of the brain would cease. The causal explanation of the movement of the individual oxygen atom seems to include a macrophysical cause—the heart's contractions. Thus, it appears to provide another example of a macrophysical cause of a microphysical effect, which (COP*) understood in terms of "microphysical" would not permit.

One might respond and say that the heart pumping blood is nothing more than the combination of more fundamental physical things such as heart chambers, composed of heart tissue, composed of cells, and so on down to the most fundamental physical particles that are structured in a particular way. This is a reasonable response, but even if this is true, it is still the structured combination of these things that performs the physiological process that causes blood to be pumped and the oxygen atom considered to be transported. And such a structured combination is not a microphysical entity.

To recap, if "physical" means microphysical, then (COP*) has several difficulties to deal with. In addition to microphysical entities, our world appears to include macrophysical entities, and there appear to be macrophysical causes of macrophysical effects as well as macrophysical causes of microphysical effects. Collectively, these issues suggest that it may not be wise to articulate (COP*) in microphysical terms.

Therefore, the physicalist may want to define "physical" in macrophysical terms. Then (COP*) would say all macrophysical events have sufficient macrophysical causes. However, as Sturgeon (1998, p. 416) points out, if "physical" means macrophysical, then the closure principle isn't supported by commonsense nor any scientific theory. Rather, "everyday experience indicates that mental events have macrophysical effects. So does macro science" (Sturgeon 1998, p. 416). Regarding the first point, that experience suggests the mental produces physical effects, we must remember two things. One, (DUAL) is plausible at the outset. In Sturgeon's (1998, p. 214) words: "Mental and physical events are distinct. This is how reality strikes us pre-theoretically." The exclusion argument aims to prove (DUAL) is false nevertheless. Yet at the outset, we cannot just presuppose that (DUAL) is false, and given (IMP) the impact of the mental on the physical, it seems plausible that irreducible mental events cause physical events. In other words, from everyday experience it seems plausible that closure is false. Moreover, as Sturgeon (1998, p. 416) points out: "No working scientific theory says broadly physical [i.e. macrophysical] effects have fully revealing broadly physical histories." In other words, science does not say macrophysical events like handshakes have merely physical causes. So if "physical" means macrophysical, closure is not supported by everyday experience nor is it scientifically supported.

At the end of the day, justification that the causal closure principle is demonstrably true requires clarity regarding what the principle actually claims. Such clarity is not only lacking, it's also difficult to secure. For one, there are various versions of the principle with different entailments. Perhaps more importantly, however, the clarity of the principle depends on the clarity of the meaning of "physical." This term is absolutely critical to the principle. Yet it's not easy for physicalists to clarify what they mean by "physical." Roadblocks crop up whether they mean *microphysical* or *macrophysical*. We have only considered challenges regarding "physical." However, another key term—"cause"—would no less require clarity and would potentially present no fewer difficulties. The physicalist owes us at least a basic clarification of what is meant by the key terms in the principle. For without such clarity, the physicalist could equivocate on the meaning of terms unknowingly. If the exclusion argument is to be logically valid, it must not equivocate on the meaning of "physical" from one premise to another.[6] In sum, to show that the causal closure principle is demonstrably true, physicalists must clarify what the principle claims, and that is a tall order.

## Inadequacy of Empirical Investigation

Clarifying the meaning of the causal closure principle is difficult to do. Nevertheless, let us assume for the sake of argument that such a prerequisite can be met. We shall now consider the merit of justifying causal closure on the basis of empirical investigation, which is a common way physicalists try to warrant the principle. In this subsection I argue that such justification fails to demonstrate that the causal closure principle is true.

To begin with, it will be helpful to consider the line of argument I have in mind. Kim (2011, p. 214) provides us with a place to start:

> Pick any physical event—say, the decay of a uranium atom or the collision of two stars in distant space—and trace its causal ancestry or posterity as far as you would like; the principle of physical causal closure says that this will never take you outside the physical domain. Thus, no causal chain involving a physical event ever crosses the boundary of the physical into the nonphysical: If $x$ is a physical event and $y$ is a cause or effect of $x$, then $y$ too must be a physical event.

Kim is not here arguing for the principle. Nonetheless, his words echo common argumentation for closure, which can be put as follows:

1. If causal closure is true, empirical investigation will lead us only to physical causes of physical effects.
2. Our empirical investigations lead us only to physical causes of physical effects.
3. Therefore, causal closure is true.

Let's call this the Empirical Investigation Argument for closure, or the E.I. Argument for short. Initially, we might think it's an argument in the form of modus ponens, which says:

A⟶B
A
∴ B

However, it is actually in the following form:

A⟶B
B
∴ A

Notice (A) does not logically follow from (A⟶B) and (B). However, that is not necessarily problematic in this case; it just means that the argument is not meant to be deductive.[7] It is either an inductive or abductive argument, which is what we would expect given that it's based on empirical observation. The conclusion of a good inductive argument is probable if the premises are true. The conclusion of a good abductive argument provides the best explanation of true premises. Given that the two argument types are similar, let's simply ask: Is the E.I. Argument a good inductive argument for closure?

It does not seem to be. For there are two good reasons to think the premises, if true, do not suggest that closure is probable. The first reason is that the majority of scientists consider it their job to find physical causes of physical phenomena. They try to do so via empirical observation. Most scientists do not see it as their job to identity nonphysical causes. The idea is that the job of the scientist, qua scientist, is to find physical causes of effects; so her theories and hypotheses are to be aimed toward that end. The empirical observations of physical scientists are not aimed at identifying nonphysical causes. Therefore, it seems inconsequential if empirical investigations of scientists do not identify nonphysical causes. After all, they are not investigating such causes nor are they investigating whether they exist. Only if one presupposes physicalism, and thus that the physical sciences give an exhaustive account of the world, does it make sense to conclude that there are no nonphysical causes because physical scientists have not identified them through their empirical investigations.

Even if scientists were looking for nonphysical causes, there is a second reason to think the premises of the E.I. Argument do not make its conclusion probable. That is, nonphysical causes, if they exist, are most likely invisible and not observable by empirical investigation alone.[8] A line of reasoning applies here which mirrors that of E.J. Lowe's (2003) in "Physical Causal Closure and the Invisibility of Mental Causation." Suppose there's a physical event $E$ that necessitates two co-causes that are always actualized together. Suppose further that one of the necessary co-causes is physical and observable while the other is nonphysical and invisible to empirical observation. Let us use "$O$" to designate the empirically observable cause and "$i$" to designate the invisible cause. To summarize, $O$ and $i$ inevitably cause $E$, and therefore when you have $O$ & $i$, you get $E$. Let us represent this idea as: ($O$&$i \longrightarrow E$).

The tricky part is that $i$ will not be recognized by the empirical investigator via empirical observations, but $O$ will be. Therefore, to the empirical observer, whenever ($O$&$i \longrightarrow E$) is true, it will appear as though ($O \longrightarrow E$) is true. Consequently, it appears that $O$ is a necessary and *sufficient* cause of $E$, but in reality it is merely necessary. Moreover, the invisible cause $i$ is also necessary and present, despite appearances. One might think we could deduce that a co-cause is necessary if there were some cases where the cause $O$ was present along with $E$, and other cases where $O$ was present and $E$ was absent. Yet, given that the two co-causes are always actualized together, such a scenario would never take place.

The basic worry is that if there were an invisible nonphysical co-cause of $E$ that was necessary, it could not be known merely on the basis of empirical investigation; nor could it be known that the empirically observable cause is not sufficient. For the empirically observable cause would appear to be the lone sufficient cause from the vantage point of empirical observation, whether or not it actually is. Note that even if such a scenario is not actual anywhere, but merely possible, it undermines the legitimacy of inferring that there are no nonphysical causes on the basis of empirical investigation. Therefore, it is impossible for empirical investigations to prove that there are no nonphysical causes (cf. BonJour 2010, p. 6; Lowe 2003).

In this subsection, the merit of the E.I. Argument as an inductive argument for closure has been challenged. I have not challenged the truthfulness of the premises meant to support the conclusion. Rather, I have given two reasons to think the premises, if true, do not make causal closure probable. For one, the empirical investigations of physical scientists are not aimed at identifying nonphysical causes; so if they don't find such causes, that doesn't mean there are no such causes. Secondly, if there are nonphysical causes, it is possible that they are invisible and unidentifiable via mere empirical investigation. What these points suggest is that the conclusion of the E.I. Argument is not supported by its premises. As a result, the argument falls short of demonstrating that closure is true. However, there is another way one might appeal to science to support closure.

*The Inevitable Success of Science*

Causal closure proponents sometimes defend their position on the basis that "research programs in physics, and the rest of the physical sciences, presuppose something like the closure principle" (Kim 2011, p. 215). This fact combined with the success of the physical sciences is thought to support the truthfulness of causal closure. The rationale goes like so: (*a*) physical scientists presuppose something akin to closure during their research; (*b*) such research has seen success that we wouldn't expect if the presupposition were false; (*c*) therefore it's likely true. Such rationale is persuasive, but several points collectively undermine it.[9]

First, the presupposition is just that—a presupposition—which scientists bring to their research, not a conclusion they derive from it. The presupposition is not a scientific hypothesis that scientists are aiming to prove by their research. So it is misguided to think that the presupposition scientists presuppose is a scientific hypothesis or theory that they have proved. That is not the case. Rather, it is an assumption they make at the outset of empirical investigation.

Second, the presupposition scientists assume is *methodological* naturalism, not *ontological* naturalism. Furthermore, methodological naturalism doesn't entail ontological naturalism. Assuming that the physical sciences are aimed at the discovery of physical causes of physical effects, scientific methodology is guided by the presupposition that there are physical causes responsible for physical effects. Given this presupposition, scientists focus their empirical investigations on discovering physical causes, and only physical causes.

Thus, it is thought that they should construct and test theories that appeal only to physical causes. The basic idea is that science investigates physical causes not nonphysical causes; therefore, the methodology of scientists should be focused on the discovery of only physical causes and assume the possibility of such discovery. But this doesn't necessarily rule out the existence of nonphysical causes. It's just that scientific methodology is not aimed at discovering such causes. So what follows, at most, from the fact that many scientists presuppose methodological naturalism is that nonphysical causes will not be discovered by scientific investigation. Yet there could still be such causes, which brings us to the next point.

Third, methodological naturalism is entirely consistent with there being nonphysical entities that have causal power to produce physical effects. Since methodological naturalism is not a presupposition about the nature of reality, but rather about the nature of scientific methodology, it does not rule out an ontology that includes nonphysical causes. Moreover, scientific success due to the presupposition of methodological naturalism would at best indicate that most effects studied by physical scientists have physical causes. That does not rule out the existence of nonphysical causes. One can endorse methodological naturalism and consistently be a dualist that believes mental causes can produce physical effects.

The renowned neuroscientist Wilder Penfield provides a prime example.[10] Penfield assumed during his experiments concerning neural impulses that the brain regions of the patients he studied were closed to nonphysical causes. Nevertheless, Penfield was a dualist (see Penfield 1975, pp. 79–82, 114). Hence, the former chair of Yale's philosophy department, Charles Hendel, fittingly comments on an early manuscript of Penfield's *The Mystery of the Mind*. In the preface, Penfield (1975, p. xii) relays Hendel's comments:

> As I read it again and again, your story is one of your starting with the physical hypothesis (accepted by all scientists as a belief in order to gain knowledge): that the physical attributes of man and energy alone are what they can deal with. You start here and cannot do otherwise, and ought not to do so. But there are discoveries made which made you wonder about something that does not fit into the scientific picture, and you wonder again and again. It is the testimony of living, conscious patients. This is an objective item in your scientific evidence. How can it be fitted into the assumed hypothesis of an entirely physical nature of man?

Penfield practiced methodological naturalism. But that didn't stop him from concluding that there is a nonphysical mind that makes decisions put into effect via the brain (see Penfield 1975, pp. 75–76). "It does this," according to Penfield, "by activating neurone-mechanisms" (Penfield 1975, p. 76).

Although Penfield presupposed methodological naturalism, no entailment of it obliged him to conclude that the physical domain is causally closed. Nonetheless, his scientific work, like the scientific field in which he worked, met with great success.

In sum, whether or not there are nonphysical causes of physical events, science has been and will be successful in providing useful knowledge about reality. That's inevitable. All it requires is the prevalence of empirically detectable physical causes that produce physical effects of interest to scientists. Such causes can be (and are) prevalent whether or not nonphysical causes exist as well. Thus, the success of science does not require the causal closure principle. So one need not conclude that the principle is true just because science is successful.

This brings us to the end of our analysis of justification for closure. Throughout this section, I have argued against justification often offered for the causal closure principle. At the outset, it was emphasized that in order to justify closure physicalists must meet the prerequisite of clarifying the closure principle, so that we can know whether or not the justification offered for it actually supports it. We have seen that meeting this prerequisite is no easy task. Never mind the various versions of the principle, clarifying the meaning of "physical" is hard enough to do in a way that does not cast doubt on the principle's veracity. I also argued against the notion that causal closure is empirically verified and the idea that the success of science supports the principle. At this point, I think it is quite safe to doubt the physicalist's claim that closure is demonstrably true and therefore dualists ought to accept it. After all, warrant commonly offered by physicalists for this claim seems to fall short.

## Reasons to Think Closure Is False

Up until this point, I've merely argued against justification offered for the causal closure principle. In light of our considerations thus far, it is reasonable to demur the claim that closure is justified and dualists ought to accept the principle. Now I will give two reasons to think closure is actually false.

*Reason one*: The initial singularity of spacetime suggests causal closure is false.[11] According to the standard Big Bang model, spacetime began at an initial singularity. This singularity either had a cause or it did not have a cause. If it had a cause, it is probable that cause was nonphysical. After all, given that space and time began at the singularity, it seems that before the singularity there would be nothing physical to cause it. Consequently, the cause would be nonphysical.

One might wish there was something physical capable of causing the singularity before spacetime began. But that would be at odds with the fact that space and time began at the singularity. For one, physical things have spatial extension, and thus a physical thing could not have existed before space. Moreover, before the singularity there was no time, and it is difficult to see how there could be a physical entity capable of causing the singularity that existed before time. To see why, consider the fact that this physical entity's causation of the singularity would be either necessary or contingent. It would either cause the singularity as a necessary outworking of its own existence or the causation of the effect would be conditional and contingent. If the former is the case, the physical entity could not have *preexisted* the effect, which would be necessary for it to exist before time given that the effect is the starting point of time. The reason the physical entity could not have preexisted the effect is simply that the cause and effect would always coexist since the effect would be a necessary corollary, or outworking, of the physical entity's existence.

Therefore, the causation of the singularity by this supposed preexistent physical entity would have to be contingent. But how could a physical entity that is outside time contingently cause the singularity? Since such a purely physical entity would not be capable of agent causation, some event would have to take place that causes the physical entity to cause the singularity. For the physical entity could not choose to cause the singularity. Consequently, something would have to take place that causes the physical entity to cause the universe. However, it seems that could not happen without there being an event before the singularity, and, therefore, time before the singularity. Although that would be inconsistent with the fact that time began at the singularity.

In the final analysis, it seems that if there was a cause of the singularity, it was probably nonphysical, and thus closure is false. Seeing what follows if the singularity had a cause, physicalists may prefer to say it had no cause at all. Yet if the singularity had no cause at all, that would undermine the commonly held version of closure that says all physical events have a sufficient physical cause. The singularity would not have a sufficient physical cause if it had no cause at all. Given this problem, one might opt for a version of the principle that doesn't require there to be any cause at all, but just says that if there is a cause, it must be physical. (COP*) is such a version; it permits the possibility of no cause but rules out nonphysical causes. Therefore, one could hold that the universe was uncaused along with the idea that science has proved causal closure. The physicalist who adopts this position could avoid surrendering causal closure by denying that there was any cause of the universe.

However, this way of salvaging closure seems to call into question the legitimacy of scientific investigation. For only if physical events have causes does it make sense to investigate what those causes are. If it's perfectly adequate to say that physical events like the initial singularity can be uncaused, what prevents that from being a viable option regarding any physical event? One might respond: "the laws of physics." But this response won't do, because the laws of physics would not place any more constraint on entities that don't exist before they pop into existence than they would on the singularity before it allegedly popped into existence. After all, before an entity exists in space and time, the laws of physics cannot affect them any more than laws of physics could have affected the singularity before it existed. The only satisfying answer, and the only answer that salvages the legitimacy of science, is that it is a metaphysical principle that physical effects have causes. So this fact doesn't depend on any other physical facts. If this metaphysical principle were not legitimate, then scientific investigations for causes would lack warrant. In the end, the route of denying that the universe had a cause undermines a metaphysical presupposition undergirding the legitimacy of science. That presupposition is: effects have sufficient causes. To deny this in order to preserve causal closure would be ad hoc and certainly more dubious than admitting nonphysical causes of physical effects.[12]

Someone might respond by claiming that the Copenhagen interpretation of quantum mechanics provides counterexamples of uncaused quantum events. I offer two points in reply. First, the Copenhagen interpretation of quantum mechanics is just one interpretation, and there are also causal interpretations, according to which quantum events have causes. For example, a theory proposed around the same time as the Copenhagen interpretation by Louis de Broglie, in 1927, provides such an alternative interpretation (Bricmont 2016, p. 129). After being neglected for about twenty-five years, the theory was proposed again by David Bohm (1952; Bohm and Hiley 1993, p. 3). The Broglie-Bohm theory, also referred to as Bohmian mechanics, provides a causal interpretation of quantum mechanics (see Goldstein 2017). It is not at all clear, at this point, that quantum events can be uncaused. Second, if there are quantum events lacking a cause identifiable by physicists, or if it empirically seems that such events do not have a cause, it does not follow that such events have no cause at all. They could have a nonphysical cause that is unidentifiable via empirical investigation and therefore appear to have no cause at all.[13] On that point, let's conclude the first reason to think closure is false and move on to a second reason.

*Reason two*: The causal closure principle undermines the possibility of rational belief in the causal closure principle (cf. Smith 2016; M. Owen 2015). By "rational belief," in this context, I mean belief that is arrived at via logical or mathematical inferences. Let's assume the exclusion arguments rationale: If closure is true, then mental causes of physical effects must be physical. Given this, a problem arises. That is, if all mental causes are physical, it would be impossible for one's belief to form as a result of logical and mathematical inferences.

For the sake of giving us an example to work with, suppose that Papineau arrived at his belief that closure is true by logically inferring it based on empirical data. Let's also suppose, however, that Plantinga arrived at his belief that closure is false by logically inferring it from the fact that God exists and caused the universe. In short, they both arrived at their beliefs rationally.

Assuming that closure is true and the rationale of the exclusion argument, it seems that neither scenario would be possible. For given such, all mental events are physical events, and therefore mental events themselves must be caused by prior physical events. But if that's the case, Papineau's belief that closure is true and Plantinga's belief that closure is false must have been caused purely by physical events. This undermines the possibility that they arrived at their beliefs rationally. To do so, they would have had to depend on inferences that are justified by laws of logic or mathematics. But such laws are nonphysical, for they could exist in nonphysical worlds and physical entities could not. As a result, laws of logic and mathematics would be excluded from playing any causal role in the production of either person's belief on the same grounds that mental causes would be excluded if they were nonphysical. So the process by which they arrive at their beliefs would not include logical or mathematical laws. Hence, neither Papineau's belief nor Plantinga's belief could be rational in the sense we stipulated above.

One might respond by claiming that the physical causal chain leading up to each person's belief would just be what a "rational process" would amount to (cf. Anscombe 1981, pp. 224–232). Thus, as the response goes, rationality would not be excluded from the process of belief production. However, such an objection misses the point: that in such a process, whether we call it a "rational process" or not, laws of logic and mathematics are excluded from playing any role in the production of the resulting belief (see M. Owen 2015). If closure were true and mental events were physical events, then the existence of every physical event would be due entirely to physical causes. Consequently, the *only* roles to be played in the production of such beliefs would be *causal roles*, and such roles would be filled by physical causes. Since laws of logic and mathematics are not physical, they couldn't play such roles. Thus, they could play no role in the production of any belief. Hence, neither Papineau's belief nor Plantinga's belief could be rational in our above sense, if closure were true. Therefore, if you have reason to believe that either Papineau or Plantinga arrived at their belief rationally, then you have reason to think closure is false.

Now we have two reasons (which do not presuppose dualism) to think closure is false. However, the dualist that already rationally believes that we have a nonphysical mind has another obvious reason to think closure is false: it seems that we mentally cause bodily effects all the time. The dualist who is justified in thinking that her mental intentionality is sui generis, distinct from its physical substrate, has strong motivation to deny closure every time it seems to her that her intention plays a causal role in her bodily actions. Additionally, many dualists are also theists. Theism and dualism fit well together, and any dualist that rationally thinks that God exists has reason to believe that closure is false.[14] Given that God would be the cause of the physical universe coming to be. Furthermore, any theist that has sufficient warrant to think God has performed miracles with physical effects has grounds for concluding that closure is false. For example, the historical evidence for the resurrection of Jesus of Nazareth could justify one's belief that closure is false, as could the evidence for contemporary miracles.[15] At the end of the day, the dualist has several reasons to think closure is false, and the theist has even further justification.

In summary, we have not only analyzed justification often given to oblige dualists to accept closure and found it wanting. We have also seen that there are reasons to think closure is false that do not depend on dualism being true. The first reason is that the initial singularity of spacetime most likely had a nonphysical cause. The second reason is that the causal closure principle undermines the possibility of rational belief in the causal closure principle itself. Therefore, it appears that dualists are fully within their intellectual rights to deny causal closure. Given that dualists can justifiably deny closure and the principle isn't entailed by their view, dualists can rationally evade the causal exclusion problem via *Route CO* without sacrificing anything essential to dualism. This is why I don't think the causal exclusion problem is dualism's chief causal problem.

## CONCLUSION

In this chapter, we have considered two other causal problems: the problem of a lack of psychophysical laws and the exclusion problem. I refer to these problems as two "other" problems because I am hereafter setting them aside. In the following chapters that deal with dualism and mental causation, these two "other" problems will not be in view. The basic reason is that I do not think they offer very strong threats to substance dualism, and I intend to deal with the problem that offers the greatest threat to dualism on causal grounds.

We have seen that the problem of a lack of psychophysical laws arises on the basis of certain assumptions, none of which are entailed by dualism. And the causal exclusion argument crucially depends on the causal closure principle, which most dualists deny, and reasonably so. In any event, it seems to me that neither of these problems pose the greatest threat to substance dualism when it comes to mental causation. The purpose of this chapter has been to clarify and distinguish these two other problems and justifiably set them aside.

We can now move forward and focus on dualism's chief causal problem—the causal pairing problem. In light of the dualist's denial of the causal closure principle, the pairing problem arises, and apart from such a denial it would not arise. The pairing problem boils down to the alleged impossibility of nonphysical causes and physical effects standing in a causal relation with one another. If legitimate, the pairing problem would certainly threaten dualism, but it would also provide the strongest a priori rationale for causal closure (cf. Kim 2011, pp. 214–215; Tiehen 2015, section 7). Given that the pairing problem is the most likely candidate for a sound a priori defense of causal closure, I didn't bother considering a priori arguments for closure above. Since the following chapter will be entirely devoted to an analysis of the causal pairing problem, which is dualism's foremost causal problem.

## NOTES

1. For helpful descriptions of event causation, see Lowe (2002, p. 157) or Plantinga (1984, p. 267).
2. John Foster (1991, p. 163) provides a clear description of the nomological assumption that informed my own articulation of it.
3. My summary is informed by Kim (2000, p. 37).
4. This formalization was guided by Robert Garcia's (2014, p. 97) presentation of the argument. David Papineau (2009) has articulated the argument in a very similar way.
5. Here, and throughout the next several pages, I am indebted to the work of Scott Sturgeon (1998).
6. Sturgeon (1998, p. 415) has shown that the "the plausibility of (COP) and (IMP) trade on distinct readings of 'physical.'"
7. If it were a deductive argument, it would clearly commit the formal fallacy sometimes referred to as affirming the consequent.
8. One would need a theoretical framework that supplies additional information in order to recognize the presence of a nonphysical cause based on empirically observable data.
9. See also Wachter (2006).
10. Stewart Goetz (2011) likewise refers to Penfield as an example, also interpreting him as a dualist.
11. My thoughts here and throughout the next several pages are informed by William Lane Craig's work on the Kalām cosmological argument. For Craig's original publication of the argument, see Craig (1979). For recent work on the argument, see Copan and Craig (2018a; 2018b).

12. Alexander Pruss (2010) makes a similar point in a brief blog post on causal closure.

13. Interestingly, in "Top-Down Causation and the Human Brain," George F. R. Ellis (2009, pp. 78, 74–76) appeals to what appears to be randomness at the quantum level to account for the fact that there are nonphysical causes of physical effects. Ellis (2015, p. 27) thinks, "The mind is not a physical entity, but it certainly is causally effective." He reasons that causal slack at the quantum level permits nonphysical causes to occur along with, and without violating, physical causation (Ellis 2009, p. 63). On his view, there are higher and lower levels of causation interacting. These levels include both physical and nonphysical causation. According to Ellis (2009, p. 78), causal interaction between the various levels is always taking place. Given such causal interaction, there may be the so-called "undetermined" events at a particular level that are actually determined by causes at another level. Therefore, in Ellis's view, if I have understood him correctly, the so-called "undetermined" quantum physical events may not be determined by physical causes, but rather by nonphysical causes.

14. Regarding the rationality of theism, see Walls and Dougherty (2018), Craig and Moreland (2009), Plantinga (2000), Moreland et al. (2013), Moreland (2008), Swinburne (1996), Copan and Craig (2018a; 2018b), and Copan and Taliaferro (2019).

15. On the historical evidence for the resurrection, see Loke (2020), Licona (2010), Wright (2003), and Swinburne (2003). Regarding contemporary miracles, see Keener (2011).

*Chapter 4*

# The Causal Pairing Problem

In the previous chapter, we saw that the causal exclusion problem arises due to the causal closure principle. Given that substance dualism in no way entails the closure principle, one way of avoiding the problem clearly presents itself to the dualist—the dualist can simply deny causal closure. Doing so costs the dualist nothing that's essential to substance dualism. And as I argued in the last chapter, dualists have warrant for such a denial. However, given such a denial, another problem appears to arise. That is, the causal pairing problem. This problem boils down to the alleged incoherence of a nonphysical mind being causally paired with physical effects.

If dualists did not deny causal closure by positing causal interaction between a nonphysical substance (the mind) and a physical substance (the body), the pairing problem would not arise. Given a denial of closure entailed by causal interaction that "mixes physical and nonphysical events in a single causal chain," it appears that dualists face the pairing problem (Kim 2000, p. 37). This problem, it seems to me, is substance dualism's chief problem regarding mental causation. Relative to the other two problems considered in chapter 3, it depends on less assumptions that dualists are disinclined to accept.

In chapter 6, I will argue that neo-Thomistic hylomorphism is a substance dualist position capable of overcoming the causal pairing problem. In that sense, I will be offering a rebuttal to the pairing problem. A prerequisite of any good rebuttal is an adequate hearing and subsequent understanding of the argument to be rebutted. Meeting this prerequisite is the aim of the current chapter. The focus of this chapter is to give a fairly in-depth presentation of the argument for the pairing problem and an analysis of its conclusion. In the section "Summarizing the Pairing Problem," a summary of the causal pairing problem will be given. The summary will be somewhat detailed, and at the end of the section a formal reconstruction of the argument for the pairing problem is presented. In the section "Analysis of Kim's Thesis," I will analyze the thesis, or conclusion, of the argument in an effort to be as clear as possible about what the charge against substance dualism is. Such clarity will aid the rebuttal offered in chapter 6.

## SUMMARIZING THE PAIRING PROBLEM

The clearest and strongest presentation of the causal pairing problem is found in the influential works of Jaegwon Kim (1934–2019). Therefore, I will focus on his presentation of the problem and argument for it in his book *Physicalism, Or Something Near Enough* and a chapter entitled "Mental Causation" published in *The Oxford Handbook of Philosophy of Mind*.

Kim (2005, pp. 73–74) tees up his argument for the pairing problem by offering an analysis of earlier arguments given by historical critics of Descartes's interactionist dualism. According to this dualist view, an immaterial mind and material body are united because they causally interact with one another (Kim 2005, p. 77).[1] Past critics include Antoine Arnauld, Pierre Gassendi, and, most famously, Princess Elisabeth of Bohemia (Kim 2009, pp. 29–30). Perhaps the first presentation of the argument is found in a letter the princess wrote to Descartes (Kim 2009, p. 31). The principal point of Descartes's objectors is straightforward: the two substances—the immaterial mind and the physical body—are of such different categories that it seems impossible for them to causally interact (see Kenny 1968, pp. 222–223).

According to Kim (2005, p. 74), however, the argumentation of Descartes's earlier critics "is incomplete and unsatisfying . . . it only expresses a vague, inchoate dissatisfaction of the sort that ought to prompt us to look for a real argument." Kim (2005, p. 74) asks, "Why is it incoherent to think that there can be causal interaction between things in 'diverse categories'? Why is it 'impossible' for things with diverse natures to enter into causal relations with one another?" Having constructively critiqued the argumentation offered by Descartes's earlier critics, Kim indicates that his aim is to go further than them and to put forth a "real argument" demonstrating why mental causation is incoherent and impossible on substance dualism (see Kim 2005, pp. 74, 86).

### Prerequisite: Pairing Relation

Kim explicitly states that Descartes's interactionist dualism is the target of his causal argument for the rejection of immaterial minds (Kim 2005, p. 71). At the outset of his argument, he invites his reader to "consider an example of physical causation" (Kim 2009, p. 32). A fundamental assumption that Kim and his interlocutors presuppose is that the nature of immaterial substances and physical substances are very different. Nonetheless, from the outset it is assumed by Kim that an example of physical causation is fittingly paradigmatic for all causation, even causation involving immaterial substances.

The example given involves two guns fired simultaneously. Gun A's firing causes X's death. Gun B's firing causes Y's death. What explains the fact that A's firing caused X's death but not Y's, and B's firing caused Y's death but not X's? Such facts cannot be unexplainable brute facts, according to Kim (2009, p. 32), who thinks, "there must be a relation R that grounds and explains the 'cause-effect pairings.'" He is quick to clarify that "we are not supposing at this point that there is a single such R for all cases of physical causation, only that some relation must ground the fact that a given cause is a cause of the particular effect that is caused by it" (Kim 2009, p. 32). Yet it does seem that Kim's example suggests one must posit an *external* relation between cause and effect. In any case, Kim's key point is that some pairing relation must ground the fact that a particular cause is paired with a particular effect.

## Necessity of Spatiality

Kim's second key point is: spatial relations pair causes with their effects (Kim 2009, p. 32). Kim considers the prospect of a causal chain fulfilling the job of linking causes with effects, but such will not suffice as an independent solution. A causal chain would only multiply relations and the fundamental question—what does the job of pairing a cause with an effect?—would remain. The answer to which, Kim believes, is spatial relations. "Intuitively, space seems to have nice causal properties," writes Kim (2009, p. 32), but he does not say whether it follows from this that only space has such properties. Nonetheless, Kim (2009, pp. 32–33) explains his intuition:

> I can state my fundamental assumption in general terms, and it is this: It is metaphysically possible for there to be two distinct physical objects, $a$ and $b$, with the same intrinsic properties and hence the same causal potential or powers; further, one of these, say $a$, causes a third object, $c$, to change in a certain way but object $b$ has no causal effect on $c$. Now, the fact that a, but not $b$, causes $c$ to change must be grounded in some fact about $a$, $b$, and $c$. Since $a$ and $b$ have the same intrinsic properties, it must be their relational properties with respect to $c$ that provide an explanation of their different causal roles vis-à-vis $c$. What relational properties, or relations, can do this job? The *only* plausible answer seems to be that it is the spatial relation between $a$ and $c$, and that between $b$ and $c$, that are responsible for the causal difference between $a$ and $b$ vis-à-vis $c$ ($a$ was in the

right spatial relation to *c*; *b* was "too far away" from c to exert any influence). At least, there is *no other possible* explanation that comes to mind.[2]

Kim has pinpointed a key prerequisite for causal relations between the purely physical entities in his examples—they stand in spatial relations. By extension, he appeals to spatial relations to explain all causal pairing relations.

Kim (2005, p. 85) also provides an explanation of how spatial relations explain causal pairing relations: "By locating each and every physical item—object and event—in an all-encompassing coordinate system, the framework of physical space imposes a determinate relation on every pair of items in that domain." Kim's account relies on the principle of the "impenetrability of matter," which says physical objects occupying the same spacetime location are one and the same object.[3] Given this principle, specific physical objects exclude other objects from a particular spatial locale (Kim 2009, pp. 34–35).

Now we could explain why the cue ball, rather than the eight ball, knocked the two ball into the corner pocket. The answer depends on their spatial locations. In sum, spatiality provides a determinate coordinate system given the impenetrability of matter, and this makes it possible for individual physical causes to be paired with specific physical effects. Since this is how causation works in the examples of physical causation he has selected, Kim (2009, p. 35) deems that causation for immaterial minds must require the same essential framework and an analogous principle of the impenetrability of matter:

> This principle is what enables space to individuate material things with identical intrinsic properties. The same goes for causation in the mental domain. What is needed to solve the pairing problem for immaterial minds is a kind of mental coordinate system, a "mental space," in which minds are each given a unique "location" at a time. Further, a principle of "impenetrability of minds" must hold in this mental coordinate system; that is, minds that occupy the same "location" in this space must be one and the same. I don't think we have any idea how a mental space of this kind could be constructed.

Given that a mental coordinate system that operates on the basis of spatial relations is necessary for causation, an obvious problem arises for immaterial, or nonphysical minds that are not spatial.

To use an example Kim (2005, p. 76) gives, suppose there are two persons, Smith and Jones, who are psychologically synchronized. Consequently, when Smith wills to raise his hand, Jones wills to raise his hand as well. In such a sorry scenario, a question arises: Why is Smith's willing causally paired with his hand rising, not Jones's, and vice versa? One might say their respective minds are united to their respective bodies and that is why each mind is causally paired with its own hand. But according to

Kim's understanding of the Cartesian's account, their minds are united to their bodies on the basis of being causally paired with their bodies. Thus, a vital conundrum arises: "The 'union' of a mind and a body that Descartes speaks of, therefore, presupposes mental causation" (Kim 2005, p. 77). It will be beneficial to elaborate on this conundrum, since it is the crux of the problem.

On Kim's understanding of Cartesian dualism, Smith's mind is causally paired with his hand rising because his mind is united to his body since his mind causes his body to move. In other words, according to Kim's reading of Descartes, the Cartesian explanation essentially says Smith's mind causes his body because his mind causes his body. This explanation is fatally circular. Furthermore, this explanation relies on a mind-body causal relation that's simply unavailable on dualism if we assume that spatiality is necessary for causality. Given that immaterial minds are nonspatial and causality requires spatiality, a causal explanation of mind-body unity is unavailable to the dualist. After all, immaterial or nonphysical minds that are not spatial cannot stand in causal pairing relations that require spatial relations. In the end, an immaterial mind cannot be united to a body via a causal relation since causation is impossible for such a mind. That's the conundrum for the Cartesian causal account of mind-body unity that Kim (2005, p. 77) thinks is the most natural option for substance dualists.

This problem might be avoided if immaterial, nonphysical Cartesian minds were somehow spatial. Yet, according to Kim (2005, pp. 88–90), supposing such would only raise more problems. For one, we allegedly lack any rational motivation for locating immaterial minds in a spatial location. Secondly, if "immaterial" souls or minds were spatial, it seems to Kim that they would just be a strange kind of matter, and therefore not immaterial. After all, on Kim's (2005, p. 88) reading of Descartes, space is the realm of matter.

In the final analysis, given their nonspatiality, immaterial minds cannot stand in causal pairing relations, whether such causal relations are mental-to-physical or mental-to-mental relations (Kim 2005, p. 87). Kim (2005, p. 87) concludes that, given dualism, "causal interaction between mind and matter is precluded by their diverse natures, and we have identified the essential diversity that matters, namely the spatiality of bodies and the supposed nonspatiality of minds." According to Kim (2005, pp. 3–4), his causal argument for the rejection of immaterial minds "shows that immaterial minds, if they existed, would be incapable of entering into any causal relations, whether with material things or with other immaterial minds." In light of the alleged impossibility of mental causation on Cartesian dualism, Kim (2005, p. 2) reasons that dualism generally "is not a workable option for anyone."

## Reconstruction of Kim's Argument

Kim does not present a formal construction of his argument in *Physicalism, Or Something Near Enough* nor in "Mental Causation" (nor any other work that I'm aware of). The following is my attempt at a formal reconstruction of his argument[4]:

(1) For every case of causation, there is a cause-effect pairing relation between cause and effect.
    1.1 For every case that $x$ caused $y$, that "$x$ caused $y$" cannot be an inexplicable brute fact.
    1.2 Therefore, for every case that $x$ caused $y$, there must be a cause-effect pairing relation that explains why $x$ caused $y$ (from 1.1).
(2) All cause-effect pairing relations require spatial relations.
    2.1 In clear cases of physical causation, spatial relations account for cause-effect pairing relations.
    2.2 Apart from spatial relations, we do not know what else could account for cause-effect pairing relations.
    2.3 Therefore, spatial relations are necessary for all cause-effect pairing relations (from 2.1 and 2.2).
(3) An entity/event must be spatial to stand in any cause-effect pairing relation (from 2).
(4) Immaterial minds are not spatial.
(5) Immaterial minds cannot stand in any cause-effect pairing relation (from 3, 4).
(6) Therefore, mental causation is impossible for immaterial minds (from 1, 5).

Now that we have laid out Kim's causal argument demonstrating the causal pairing problem for dualism, let's analyze Kim's thesis, the conclusion of the argument.

## ANALYSIS OF KIM'S THESIS

Kim's argument is for the conclusion that it is impossible for an immaterial, nonphysical mind or soul to cause something, not merely that none in fact do (Bailey et al. 2011, p. 351). Why opt for this stronger claim? To rule dualism out, or to warrant a complete rejection of immaterial minds, it must be shown that mental causation is impossible on dualism, not just mysterious uncharted territory. Otherwise, if there were good reasons to believe (a) that humans are composed of a body and an immaterial mind and (b) that such a mind is causally responsible for some physical states of our bodies, we could justifiably conclude (c) that there is genuine top-down mental causation, *whether or not we understand such causation.*[5] As long as such causation is possible and we have good reasons to think there is such causation, there is no warrant for the rejection of dualism on causal grounds. We would simply be faced with territory yet to be understood.

Since Kim's aim is to justify a rejection of immaterial minds, his argumentation is judicious. His thesis—mental causation is *impossible* for immaterial minds—may seem too ambitious to some, yet it's needed to warrant his rejection of dualism. Yet, we must ask: impossible, in what sense? Kim does not say in what sense mental causation is impossible for nonphysical minds. The reader is left to figure this out.

There are four relevant senses of impossibility. Something is physically impossible if the laws of physics preclude it from happening.[6] It is physically impossible for an elephant to fly like a bird. Something is logically impossible if the laws of logic preclude it from ever being so. To use a popular example, it is logically impossible for there to be a married bachelor. For given the meaning of the terms "married" and "bachelor," the law of noncontradiction precludes the possibility of married bachelors.

The third sense of impossibility is nomological impossibility, which is like physical impossibility since it pertains to laws. Assuming there are natural laws beyond the laws of physics, something is nomologically impossible if such laws preclude it. The fourth relevant sense of impossibility is metaphysical impossibility. Something is metaphysically impossible if it cannot be given the natures of the entities in question (cf. Moreland and Craig 2003, p. 50). It is metaphysically impossible for two rocks to fall helplessly in love. Since it is essential to the nature of a relationship of love that the entities standing in the relation be personal (let's assume), but by nature rocks are impersonal. Thus, it is metaphysically impossible for two rocks to be in love. Let us now try to decipher what sense of "impossibility" is likely meant in Kim's conclusion that mental causation is impossible for immaterial minds.

## Physically Impossible?

The claim that mental causation on dualism is physically impossible would be as interesting as a flat, straight, road in the middle of Kansas cornfields. That is, not interesting at all, nor would such be consequential. For if there were nonphysical minds that have causal power to affect nonphysical mental states and physical bodily states, we have no reason to suspect that such causation would be purely physical and therefore governed only by principles governing merely physical entities. For the entities doing the causing *ex hypothesis* would not themselves be merely physical, nor would all the effects be purely physical. Apart from the assumption that physicalism is true, we couldn't justifiably expect that such causation would necessarily be purely physical or governed merely by laws of physics. Rather, given the nonphysical nature of the cause, we could expect the causation to be supra-physical. In other words, we could reasonably expect the governing principles of such causation to go beyond the laws of physics, while not necessarily violating them. Such principles could be in concert with physical laws and yet not be identical to physical laws, nor regulated by them.[7]

The dualist could grant that such causation is not physically possible in the sense that laws of physics alone could govern it. Yet, she would do well to point out that the primary question is whether or not there could be a coherent framework that includes principles of causation beyond just the laws of physics that govern causation involving nonphysical entities. To answer this question by asserting that such is physically impossible is inadequate, for the question itself is about whether or not something beyond mere physical possibility is possible. If Kim's conclusion was that mental causation is physically impossible for immaterial minds, the primary question would remain unanswered. Consequently, Kim's justification for the rejection of immaterial minds would be lacking.

## Logically Impossible?

Given that a physical impossibility of mental causation on dualism would be inconsequential, one might think that Kim's conclusion pertains to logical impossibility. However, if the conclusion of Kim's argument is that mental causation is logically impossible on dualism, his premises fall short of guaranteeing such a conclusion. Consider this simplified formulation of his argument:

(*P1*) Necessarily, spatiality is needed for causality.
(*P2*) Necessarily, immaterial minds are not spatial.
(*C*) Necessarily, mental causation is impossible for immaterial minds.

For the impossibility arrived at in *C* to be a logical impossibility, the necessity of *C* must be of logical necessity. Yet, if the necessity of *C* is a logical necessity and *C* is to follow from *P1* and *P2*, the necessities of both *P1* and *P2* must be logical necessities as well. And for these necessities to be logical necessities, the propositions they modify must be entailed by the meaning of the relevant terms, that is "causation" and "immaterial minds."

Therefore, in order for the necessity of *P1* to be one of logical necessity, it must follow logically from the meaning of "causality" that spatiality is needed. Kim offers his reader rationale for why he thinks causation in the actual world requires spatiality. He does so through giving examples of causation in the actual world and pointing out how causation works in those cases. But showing that causation is a certain way in the actual world is not the same as showing that causation could not have been otherwise out of logical necessity. And that is what must be done in order to show that Kim's concept of causation in the actual world is one that's logically necessary in any possible world. Thus, even if Kim's depiction of causation in the actual world was perfectly accurate, it would not prove that his understanding of causation and what it requires is logically true. So if Kim's conclusion is of logical necessity, his argument is severely weakened by the fact that he only offers justification to think causation requires spatiality in the actual world. He needs to demonstrate that causality logically requires spatiality in order for his argument to be sound. At best, Kim has shown that, given the laws of physics in our world, cases of mere physical causation require spatiality.

Not only does Kim's argumentation lack rationale justifying the idea that causality logically necessitates spatiality, such an idea is doubtfully true. It seems logically possible for there to be different laws governing causation with different prerequisites than what the laws of physics require. For if this were not so, empirical investigation would be unnecessary for discovering truths about laws of physics. After all, if such truths were logically necessary, we could arrive at such truths through a priori reasoning *alone*. Yet the discovery of physical laws often depends on empirical investigation, even if a priori reasoning is involved. So not only does Kim's argumentation lack justification for thinking *P1* is true in the sense of logical necessity, it is doubtfully true in that sense.

Yet, even if it were true that causality requires spatiality out of logical necessity, it is doubtful that such is demonstrably true. In the case that causality logically necessitated spatiality, such a truth would clearly undermine some of the most widely held views throughout human history and during contemporary times. And given that such a truth would be knowable a priori, since it would follow logically from the meaning of "causation," it would be dumbfounding that it escaped the notice of countless thinkers. After all, many thinkers throughout Western civilization have thought it is at least logically coherent to think an immaterial, nonspatial being caused the spacetime realm to come into existence and to persist with consistency. Many Eastern thinkers have thought that the idea of Karma being causally responsible for certain events in our world is logically coherent.

However, causality logically necessitating spatiality would not only be at odds with religious ideology. As Andrew Bailey, Joshua Rasmussen, and Luke Van Horn (2011, p. 351) have pointed out, along with J. P. Moreland (2005), the idea that causality requires spatiality entails that the singularity on the standard model of the Big Bang could not have a cause. According to the model, the entire spacetime continuum began at the singularity. Therefore, if causality necessitates spatiality, then any model proposing that the Big Bang had any cause at all—such as Quentin Smith's (2002) atheistic explanation of the singularity—would be disqualified.

That would be odd. The idea that there can be effects like the Big Bang that have no cause whatsoever seems far more incoherent than the idea that there can be nonspatial causes of such effects.[8] After all, I think C. D. Broad (1955, p. 10) does not merely speak for himself when he says: "I cannot really *believe in* anything beginning to exist without being *caused* (in the old-fashioned sense of *produced* or *generated*) by something else which existed before and up to the moment when the thing in question began to exist." In the case of the initial singularity, there is much disagreement about the nature of the cause, yet it has seemed quite reasonable to many to suppose it had a cause.

In sum, Kim has not given us reasons to believe that it is logically necessary that spatiality is required for causality, which is needed for his argument to be sound if *C* is of logical necessity. To boot, it's likely false that causality requires spatiality out of logical necessity. Even if such were true, it is doubtful that it's demonstrably true. Given that numerous thinkers have held beliefs that clearly stand in opposition to such a "fact" that would be knowable a priori. Hence, it is safe to say with physicalist, David Papineau (2009, p. 55), "there is nothing conceptually contradictory in the idea that physical phenomena may be affected by non-physical causes, as Descartes supposed, for example." Nonetheless, there are more senses of impossibility.

## Nomologically Impossible?

Nomological impossibility is the third sense of impossibility worth considering. Something is nomologically impossible if it cannot be due to laws of nature, whether such laws are laws of physics or another type of natural law that might even be nonphysical.

Due to biological laws (let's assume), it is possible to cut a limb from one apple tree that grows Red Delicious apples and graft it into another apple tree that grows Granny Smith apples, so the tree grows apples of both types. This is an example of a nomological possibility. Likewise, it is nomologically impossible to graft the same apple tree limb into a pine tree so that it grows both pinecones and Red Delicious apples.[9]

To elucidate how nomological impossibility could apply to dualism and mental causation, let us imagine that the natural laws in our universe include psychophysical laws that govern all causation between the mental and the physical. Suppose that these laws permit physical causes of nonphysical mental effects but preclude nonphysical mental causes of physical effects. If this were the case, it would be nomologically impossible for there to be nonphysical mental causes of physical effects in the body. And there would be no laws that could causally pair mental causes with physical effects, if such laws were needed.

As one reads the earliest works related to the causal pairing problem, one might get the impression that nomological impossibility is the issue.[10] However, Kim's most developed case for the causal pairing problem, which I've focused on in this chapter, does not appeal to laws in order to demonstrate that mental causation is impossible for nonphysical minds (see Kim 2005, Ch. 3). Rather, he appeals to natures, which strengthens his case.

If Kim's argument hinged on an appeal to laws, there would be several ways the dualist might try to escape the conclusion. The dualist might (1) deny that such laws actually govern causation, (2) deny that such laws govern all causation, (3) deny the existence of the laws that purportedly preclude nonphysical minds from producing physical effects, or (4) only accept revised versions of the laws that don't preclude such mental causation (cf. John Foster 1991, Ch. 6). However, these escape routes are unavailable if the causal pairing problem is not rooted in laws but rather in the very nature of causation, physical bodies, and nonphysical minds.

So it is not surprising that Kim's (2005, Ch. 3) most developed presentation of the argument for the causal pairing problem does not appeal to laws, but rather to natures.[11] His emphasis on natures—the nature of causation which relies on spatial relations, the spatial nature of physical bodies, and the nonspatial nature of nonphysical minds—strongly suggests that metaphysical impossibility is the type of impossibility Kim's conclusion pertains to (as I will discuss in the following section).

Those who think laws of nature are grounded in natures might object as follows. When we consider the above example regarding biological laws and grafting an apple tree limb, biological *laws* grounded in the natures of the apple tree and pine tree are what make one scenario possible and the other impossible. The nature of the apple tree entails certain laws that make it possible to graft the limb, and the nature of the pine tree entails certain laws that make it impossible to graft the limb. The possibility and impossibility are determined by the laws, which are entailed by the relevant natures. The same is true with regards to the impossibility that the causal pairing problem presents, one might argue. It is laws grounded in the natures of causation, physical bodies, and nonphysical minds that make mental causation impossible for nonphysical minds. Therefore, the causal pairing problem identifies a nomological impossibility based on laws, which are grounded in natures.

The idea that laws of nature are grounded in natures is reasonable. But if this is so, the laws grounded in natures are less fundamental than the natures that ground them. And if the laws responsible for the causal pairing problem are entailed by the natures in question, it seems that the root of the problem is the natures. Thus, if Kim's argument shows that the very nature of causation, as well as the natures of physical bodies and nonphysical minds, prompts the causal pairing problem, he will have identified the fundamental root of the problem. Additionally, his argument will close off the escape routes mentioned above. In short, his argument will be stronger, and the problem identified will be more threatening to dualism. In light of this, I think our fourth sense of impossibility that pertains to natures—metaphysical impossibility—provides the best interpretation of Kim's argument and its conclusion.

## Metaphysically Impossible?

To say that mental causation is metaphysically impossible for immaterial minds is to make a stronger and more substantive claim than saying such causation is physically impossible. Yet it's a more conservative claim than saying that such causation is logically impossible. As we have seen, if Kim's thesis pertains to physical impossibility it's inconsequential, but if his thesis pertains to logical impossibility, then his premises don't warrant such a conclusion. In any event, Kim need not worry about such charges. For it seems that his thesis pertains to neither sense of impossibility, but rather to metaphysical impossibility.[12]

As mentioned above, something is metaphysically impossible if the natures of the entities in question preclude it from happening. While setting up his argument, Kim (2005, p. 74) asks: Why is it impossible for entities of *different natures* to stand in causal relations with one another? Moreover, Kim (2005, p. 87) concludes that causal interaction between an immaterial mind and matter "is precluded by their diverse natures." And Kim (2005, p. 87) suggests that he has pinpointed the diversity that precludes such—that is, the nonspatiality of immaterial minds and the spatiality of material bodies. This is so only if the nature of causality requires spatiality, which Kim thinks it does. Thus, the nature of immaterial minds and the nature of causality are thought to give rise to the causal pairing problem. This problem amounts to the alleged metaphysical impossibility of a nonphysical mind causing a physical effect, or, in other words, the metaphysical impossibility of mental causation on dualism.

In closing, let us summarize the above analysis. Kim's thesis regarding the causal pairing problem is not that mental causation is physically, logically, or nomologically impossible for nonphysical minds. Rather, it is that mental causation is metaphysically impossible. The idea is that the nonspatial nature of immaterial or nonphysical minds and the nature of causation, which supposedly requires spatial relations, would preclude such minds from causing effects. In other words, causality requires spatiality but immaterial minds are not spatial, and thus mental causation on dualism is allegedly incoherent.

## CONCLUSION

The greatest objection to dualism regarding mental causation is the causal pairing problem. As we saw in the previous chapter, the problem of a lack of psychophysical laws rests on several assumptions dualists need not accept and the causal exclusion problem hinges on a premise dualism does not entail in any way. The causal pairing problem, however, is aimed at objecting to dualism on grounds dualists are more inclined to accept.

Since the seventeenth century, a mechanistic view of causation has become increasingly favorable, and it is often thought to depend on spatial relations. Yet according to the most well-known substance dualist view, Cartesian dualism, nonphysical minds are nonspatial. Consequently, a framework of mental causation including a mechanistic view of causation and nonphysical minds appears incoherent, as the pairing problem suggests. The apparent inconsistency is simply that causation requires spatiality but minds are not spatial. Hence, the pairing problem allegedly undermines the very coherence of dualism with respect to mental causation. Therefore, the causal pairing problem presents a grave potential defeater of dualism.

In chapter 6, I will argue that neo-Thomistic hylomorphism can defeat this potential defeater. Although before applying the view to the pairing problem, I will introduce it in chapter 5.

## NOTES

1. Although Kim (2005, p. 77) appears to express some doubt about the historical accuracy of his reading of Descartes's view on mind-body unity (cf. Stump 2003, p. 512).

2. Emphasis mine.

3. One might ask: What about coincident objects like a lump of clay and a statue? Kim acknowledges the issue of coincident objects and understandably sets it aside, not to be distracted from the main trajectory of his argument (see Kim 2009, p. 34, footnote 6).

4. For another formulation of Kim's argument, and a formulation of a more modest version of the "Pairing Argument," see Andrew M. Bailey et al. (2011).

5. Ellis (2009, p. 75) makes a similar point; and Swinburne (2013, p. 105) also makes a very similar point, if not the same point.

6. For a brief discussion of physical necessity, the corollary of physical impossibility, see Chisholm (1976, p. 22).

7. Cf. Ellis (2009).

8. See also chapter 3, the section "Reasons to Think Closure is False."

9. Or, we can at least assume that this is the case for the sake of the example. I must admit I have not tested the claim myself, nor do I know of anyone who has.

10. According to Kim (2001a, p. 35, footnote 7), his thinking on the causal pairing problem was prompted by Foster's (1968) article "Psychophysical Causal Relations," and Kim's (1973) first discussion of the causal pairing problem is in an article entitled "Causation, Nomic Subsumption, and the Concept of Event."

11. Kim (2005, p. 76) seems clearly interested in closing off escape route (1).

12. Bailey et al. (2011, p. 350, footnote 3) likewise interpret Kim.

*Chapter 5*

# Neo-Thomistic Hylomorphism

"Cartesian dualism has clear and unassailable pride of place as the whipping post on which dualists are ritualistically flailed," notes David Oderberg (2005, p. 71). When it comes to considering and critiquing substance dualism, a depiction of Descartes's view is almost always at center stage. At this point, we will diverge from such orthopraxy as we focus on a different form of substance dualism—neo-Thomistic hylomorphism.

The aim of this chapter is to summarize neo-Thomistic hylomorphism, which I will often refer to simply as "hylomorphism" for brevity.[1] The following two chapters will apply the view to the causal pairing problem and accounting for neural correlates of consciousness. While these subsequent chapters will deepen our understanding of hylomorphism, this chapter is introductory. The goal is merely to clarify the view.

I have labeled the view "neo-Thomistic hylomorphism" because I think it is appropriate to acknowledge Aquinas's influence. Since the fundamental principles of the view are derived from his thoughts on human ontology, which were greatly shaped by Aristotle's works. Yet, for several reasons I call it *neo*-Thomistic hylomorphism. For one, I make no claim that the view is exactly what Aquinas held. Although my neo-Thomistic view is significantly influenced by his ideas about human nature, I give myself the freedom to disagree with Aquinas on certain points or to modify his ideas in order to make the view more defensible. I look to Aquinas as a philosophical authority who provides a good starting point, but not as an infallible final word. Furthermore, my focus is not on exegeting Aquinas's works. My intention is not to engage in hermeneutical debates about how to interpret Aquinas.

As valuable as such debates are, they are not my focus in this particular work, and I am aware that other philosophers have different interpretations of Aquinas than I do on key points.[2] My focus is rather on applying a view that is informed by Aquinas to chief contemporary problems, specifically those that confront substance dualism. In the end, I hope to have shown that neo-Thomistic hylomorphism can adequately handle such problems.

While presenting a view of the mind and body is the purpose of this chapter, it would be naïve to think such could be done in a metaphysical vacuum. For Aristotelians, such as Aquinas, metaphysics is the "engine that drives every other field of philosophy" (Oderberg 1999, p. ix). I do wish to propose a "mere-hylomorphic" view in the sense that one can adopt the mind-body view presented without becoming a full-blown Aristotelian or Thomist on every philosophical point. Yet I do not wish to propose a "shallow-hylomorphic" view in the sense that hylomorphism is presented in a context devoid of metaphysical presuppositions that are critical to the view. Hence, our considerations of hylomorphism will involve metaphysical tenets importantly related to, but not necessarily proper to, the philosophy of mind.

In the section "Aristotelian Terminology," we'll wade into the waters of Aristotelian-Thomistic metaphysics in order to introduce key terms and concepts. The following section, "Human Ontology," will focus on Aquinas's human ontology and the nature of human persons. Then the section "Incarnate Souls" will specifically discuss the human soul. Our considerations of the soul will move from focusing on the soul's relation to the body to clarify the nature of the soul in itself. Once the nature of the soul is clarified, it will become apparent that the hylomorphic view I am proposing is a substance dualist position. The concluding section will offer a summary of neo-Thomistic hylomorphism.

## ARISTOTELIAN TERMINOLOGY

I wish we could just translate all Aristotelian terms into our contemporary vernacular. However, some Aristotelian terms convey specific ideas not entirely captured by contemporary metaphysical terms. Therefore, it is worth clarifying some essentials before trying to articulate neo-Thomistic hylomorphism. This section is devoted to doing just that. I won't cover all the terms that will subsequently be employed. But I will introduce enough terms to start us down the path of explicating hylomorphism. The systems of thought referred to as "Aristotelianism," "Thomism," and "Scholasticism" have benefited from some of the longest time periods of intellectual consensus and tradition in Western culture. As a result, these systems of thought, which share in common their Aristotelian elements, are very substantive as well as nuanced. I will merely be scratching the surface of an extraordinarily rich intellectual tradition.

Nevertheless, one must begin somewhere, and I begin with clarifying the term *hylomorphism*. Then substances will be distinguished from aggregates before we clarify the Aristotelian notion of a *form*. The concept of *matter* will then be discussed. At the end of this section, we will employ the Aristotelian terms already discussed to clarify what a material substance is and to introduce what I call an "en-forming relation."

## Hylomorphism

Fittingly, "hylomorphism" is a compound Greek word that conveys the idea of a composite of form and matter. The Greek word "hyle" means *matter*, while the word "morphe" means *form*. Put the two together and you get "hylomorphism." Christopher Shields (2014) offers the following definition:

Hylomorphism $=_{df}$ ordinary objects are composites of matter and form.

As the words "hyle" and "morphe" together amount to "hylomorphism," a real form and matter together constitute a concrete hylomorphic object. The key point expressed by this basic definition is that objects have two constituents: form and matter. While this definition is helpful, it's quite basic, and in a sense too basic. For according to an Aristotelian-Thomistic ontology, there are aggregates and then there are substances. The constitution of the former is significantly different than that of the latter.

## Substance versus Aggregate

A pile of logs is an example of an aggregate, and a tiger is an example of a substance from the standpoint of Aristotelian metaphysics. In this section, substances will be distinguished from aggregates.

My discussion will focus on three differences. First, substances are more ontologically fundamental than aggregates since aggregates are composed of substances, whereas substances are not composed of further substances (see Inman 2018, pp. 102–106). Second, the parts of a substance (if it has parts) only have existence as a part of the whole substance, whereas the parts of an aggregate can have existence that's independent of the whole aggregate. Third, and most importantly, substances have an internal unity that aggregates lack (cf. Marmodoro and Page 2016, pp. 6–7). This is because one single form grounds the existence and essence of a substance and all its parts. By contrast, the parts of an aggregate are themselves substances with their own individual forms that ground their own individual existence as the kind of substance they are. So an aggregate depends on the forms that ground the existence of its parts, which are substances, as well as the form that modifies the parts in a way that unites the parts into one aggregate.

Let's unpack these differences, beginning with the fundamentality of substances vis-à-vis aggregates. Substances are metaphysically prior to aggregates. Put differently, substances are more ontologically fundamental than aggregates. The reason is that aggregates are composed of substances, and therefore substances are more basic than what they compose. Aggregates depend on the existence of their constituent substances like a pile of rocks depends on the rocks composing the pile. If you take away the rocks, the pile no longer exists. The pile's existence depends on the existence of the rocks. Likewise, any aggregate's existence depends on the existence of the substances that are parts of the aggregate.

This brings us to the second difference, which pertains to the parts of a substance versus aggregates. Briefly stated, the parts of a substance are not themselves substances, and thus cannot exist apart from the whole substance they are a part of, but this is not the case with aggregates. The inseparable parts of substances are not themselves substances, and thus do not have their own existence apart from the substance they're a part of. However, the separable parts of an aggregate that are themselves substances have their own existence, and thus can exist apart from the whole aggregate they're part of.

To contrast the inseparable parts of a substance with the separable parts of an aggregate that are ontologically prior to the aggregate, imagine a human pyramid composed of Socrates, Parmenides, and Pythagoras. When Parmenides and Pythagoras crouch down and Socrates climbs upon their backs, a pyramid-shaped aggregate comes into existence—call it *the Philosophers Pyramid*. While the existence of the Philosophers Pyramid depends upon the existence of the three philosophers and the configuration consisting of relations they stand in to one another, the existence of each philosopher does not depend on the existence of the Philosophers Pyramid. Before it existed, they existed. After it ceases to be, when Parmenides and Pythagoras get tired of holding up Socrates, they will still exist. This is because they themselves are substances that can exist apart from the whole—that is, the Philosophers Pyramid—that they once were a part of. By contrast, assuming that Socrates's body is a substance, his right hand is an inseparable part of his body. If you remove Socrates's hand from the whole of Socrates's body of which it is a part, it will cease to be Socrates's hand. Granted, we could still loosely call it "Socrates's hand," but strictly speaking, it is no longer Socrates's hand since it is no longer part of the substance it was an inseparable part of.

The substances that are the parts of aggregates are separable parts because their existence does not depend on being a part of the aggregate. They have their own existence apart from the aggregate—that is, apart from the whole. By contrast, the parts of a substance are inseparable parts that do not and cannot exist apart from the substance they're part of. The reason why the separable parts of aggregates have their own existence apart from the whole aggregate they're part of and that the parts of substances do not have existence apart from the whole substance they're part of pertains to the difference of unity aggregates have versus the unity substances have. An aggregate consists of multiple substances that are its parts, and each of these substances has its own individual form that grounds its existence as the kind of thing it is. So each substance that is part of an aggregate has existence in virtue of its own individual form.

The parts of a substance are very different. One form grounds the existence and essence of the whole substance including each of its parts. Such a form is called a "substantial form" (see the section "Forms"). A single substantial form grounds the existence and essence of the whole substance that it is the form of, including all its parts. In other words, one substantial form grounds the existence of a substance along with all the parts of the substance it enforms. The parts of the substance are unified in virtue of the existence and essence of each part being grounded by one and the same form. Thus, a substance is unified to the highest degree as each of its parts shares the same form and exists only as a part of the substance they're a part of (see Inman 2018, pp. 103–104).

A human body is a common example of a substance. And as will be discussed further below, Aristotle and Aquinas thought the soul is the substantial form of the body, which grounds the body's existence by grounding its unity and essence. Socrates's body is a unified whole body since the existence and essence of each of its parts is grounded by one individual form, which is Socrates's soul. Unlike the parts of aggregates, Socrates's hand cannot exist apart from the whole substance and the form that unifies it. The parts of substances are inseparable parts. As mentioned above, if Socrates's hand were severed from his body, it would no longer literally be what it once was, according to an Aristotelian understanding of substances and their parts. We could debate what it would be, but one thing it would not be is Socrates's hand.

We can now recap the differences here considered between substances and aggregates. To begin with, an aggregate is an entity composed of multiple substances. The substantial form of each substance that's part of an aggregate grounds its individual existence. The existence of the aggregate depends on the substances that are its parts, and each of these substances depends on its own individual substantial form. Therefore, the aggregate's existence depends on multiple substantial forms grounding the existence of each part.

In contrast, a single substantial form grounds the existence of a whole substance including its parts. Therefore, a substance is fundamental in that it is not composed of further substances with their own individual substantial forms. The parts of a substance have their existence and their essence grounded by the same form that grounds the existence and essence of the whole substance. This is why a substance is internally united in a way that aggregates are not. Unlike the inseparable parts of a substance, the parts of an aggregate do not depend on the same form for their existence and essence.

There are two passages from Aquinas's writings that capture especially well the foundational differences between substances and aggregates that we have here discussed. As we conclude our discussion of substances and aggregates, it will be instructive to briefly consider these passages.

In each passage, Aquinas uses the example of a body unified by a substantial form and a house composed of substances unified by a configuration that is an "accidental form"[3]:

> Now the substantial form perfects not only the whole, but each part of the whole. For since a whole consists of parts, a form of the whole which does not give existence to each of the parts of the body, is a form consisting in composition and order, such as the form of a house; and such a form is accidental. But the soul is a substantial form; and therefore it must be the form and the act, not only of the whole, but also of each part. (Aquinas, ST 1a 76.8c)[4]

The substantial form of the body grounds the existence of every part of the body. By contrast, the parts of a house, which are themselves substances, have their existence and essence grounded in their own distinct substantial forms. The substances composing the house are then united by an accidental form, which is the ordered configuration they are in. As a result, a distinguishing feature of a substance is its oneness, or internal unity, due to the one form that unites it by grounding the existence of all its parts:

> For since the body of a man or that of any other animal is a certain natural whole, it will be said to be *one* because it has *one form* whereby it is perfected, and not simply because it is an aggregate or a composition, as occurs in the case of a house and other things of this kind. (Aquinas, QDA 10c in Marmodoro and Page 2016, p. 12)[5]

The body is internally unified because one form grounds the existence and essence of the whole substance including all its parts. The house, by contrast, consists of substances that have their own substantial forms. These substances, which are the parts of the house, are externally unified by an accidental form, which is the ordered configuration they are in. At this point, it behooves us to discuss forms in more depth.

## Forms

To understand forms, it is important to know that there are both accidental forms and substantial forms. An accidental form modifies an existing entity (or entities), whereas a substantial form grounds the existence of a whole substance including its parts. A substantial form grounds the existence of a substance in virtue of grounding its unity and essence; thus, it also grounds the properties and powers the substance has and naturally develops due to its essence.[6] In this section, I will briefly describe accidental forms and then elaborate on substantial forms.

The shape of a statue is a well-known example of a form. When most people think of an Aristotelian form, they think of a shape. Yet, if a shape is all that comes to mind when one considers the hylomorphist's claim that the soul is the form of the body, misunderstanding will abound. A shape is an accidental form, whereas the soul is a substantial form.

Further examples of accidental forms are the whiteness of a bench, the temperature of a body of water, and the firmness of a mango. Notice that these forms do not ground the existence of what they modify. The bench could become pink and still exist. The water could change its temperature without ceasing to exist. The mango could become soft as it ripens while continuing in existence. Each entity could exist without the specified accidental form. After all, such a form does not ground the entity's existence. Moreover, the bench could be *what* it is—a bench—with a different color. The water could be water if its temperature changed. The mango could still be a mango if it softened as it ripened. The specified forms do not ground the essence of the entities in question any more than they ground the existence of the entities.

Unlike accidental forms, a substantial form grounds the existence of what it en-forms (see the section "En-forming Relation" on "en-forming"). A substantial form grounds the existence of a substance that it is the substantial form of in virtue of grounding its unity and essence (see Marmodoro and Page 2016, p. 15). This has multiple relevant implications. First, since the substantial form grounds the unity of the substance by en-forming every part of the substance, the substantial form grounds not only the existence and essence of the substance but also its parts. Second, since the substantial form grounds the essence of the substance and its parts, the substantial form is ultimately what grounds what properties and powers (or capacities/dispositions) the substance and its parts essentially have and develop. For clarification, a substance and its parts can also have nonessential properties that are not grounded by the substantial form (e.g., a tiger's spatial location is a nonessential property that is not grounded in its form).

To further elucidate the idea of a substantial form, let's consider an apple tree, assuming that such is a substance. Suppose that in the summer of 2020 I planted a very young apple tree in my yard, to pass the time during the coronavirus pandemic. Suppose further that the tree was so young that it did not have limbs or leaves when first planted, and therefore no fruit. Despite this, I know that the tree will come to have a structure with limbs and leaves in due course, as long as the tree is well kept and maintains good health. Moreover, if it stays healthy for several years it will develop to the point where it consistently produces a fruit with a very standard structure. The structure of this fruit is so standard and predictable that we can imagine the shape of the fruit simply by knowing that the tree is an apple tree.

About five years after being planted, the tree will predictably develop certain parts—limbs, leaves, and apples—with standard structures. While the particular structures of each of the limbs can vary to some degree, their structures will be standard enough that one could easily discern the apple tree limbs from willow tree limbs, or pine tree limbs, and so on. The apple tree will develop these parts with certain structures due to the kind of thing it is, which depends on the form that grounds its essence. Because the physical material of the apple tree has the form of an apple tree, it will develop to have apple tree limbs, leaves, and apples by the summer of 2025.

But let us entertain the idea that in the autumn of 2025 the leaves and apples fall off the tree, and deer chew off all its young branches. Even so, there's no need to fret. Assuming I maintain the tree well, it will regrow limbs, leaves, and apples over time. The old leaves and apples that grew in the summer of 2025 and fell off that autumn will decompose. The limbs eaten by deer will be digested and return to the soil. At the end of this cycle, the apple tree will have new limbs, leaves, and apples. Although these new parts will be composed of different matter than the original parts.

Since an apple tree can live for a hundred years, this process of what we might call "part regrowth" can happen multiple times. The tree can lose its old parts and then regrow lost parts with different constituent matter. By the time the tree is 100 years old, it will have changed significantly, and the matter that constitutes it will be different from the matter that constituted the tree when it was first planted. Nevertheless, the tree will persist. Despite the change, the tree that I planted in the year 2020 can live to 2120. That which makes it the same tree throughout all those years, and all the changes it undergoes including part regrowth, is its substantial form. Since, from 2020 to 2120, the tree has the same substantial form that grounds its existence, it is the same tree throughout the time and change. Thus, the substantial form also provides a way of explaining how the tree can persist through significant time and change. The tree persists because its substantial form persists through the time and change.

This is fitting since the substantial form of the apple tree is what grounds the existence of the tree as a unified apple tree at any given time. Since every material constituent of the tree shares the same form that grounds its existence, every material constituent is united as a material constituent of the same organism, the apple tree. Moreover, the apple tree is an apple tree, rather than say a dinosaur, because of the substantial form it has. If it had the substantial form of a dinosaur, it would have different properties, different powers, and different parts with different structures capable of manifesting its different powers. Simply put, if the entity had the substantial form of a dinosaur rather than that of an apple tree, it would be a dinosaur with the characteristics of a dinosaur, not an apple tree.

In the final analysis the substantial form provides the ultimate natural explanation for the substance's existence.[7] In the case of our apple tree, the substantial form grounds the existence of the tree by grounding its unity as well as its essence, and therefore its essential properties and powers. And the parts of the apple tree are en-formed by the same substantial form, which therefore grounds their existence and essence as well. Given that the form grounds the tree's existence and that of its parts, we could say the tree and its parts ontologically depend on the form.

By way of summary, according to Aristotelian metaphysics there are accidental forms and substantial forms. Accidental forms modify that which they are the accidental form of, and an entity may have multiple accidental forms that modify it. By contrast, a substance will have only one substantial form that grounds its existence by grounding its unity and essence. Now that we have a basic grasp of forms, we should consider the other constituent of a hylomorphic object.

## Matter

Matter is the physical material of an entity. The matter of a ball is the physical material comprising the round ball. The matter of my wedding ring is the physical material wrapped around my finger. The matter of the statue *David* is the physical material shaped like the man, King David.

Notice that matter is the physical constituent *of* an entity, on Aristotelianism. All matter is the matter of something. Since all matter is the physical constituent of something that exists, it too exists. While Aristotelian and Thomistic metaphysics includes the concept of "prime matter," such matter doesn't actually exist.[8] "Prime matter" refers to matter that has no form. But all matter that exists has a form(s), according to Aristotelianism. So prime matter does not actually exist. Matter, on the other hand, does exist, and it is a constituent of a particular type of substance—a material substance.

## Material Substance

A material substance is a substance that consists of a substantial form that naturally en-forms matter. Human persons, according to Aquinas, are material substances given that the human soul en-forms matter, constituting a physical body. By contrast, bodiless angels are immaterial substances. The difference between a material substance and an immaterial substance is that the substantial form of a material substance en-forms matter, which is not true of immaterial substances. Given that I have already used and will continue using the term "en-form," it will be helpful to clarify it and the related concept "en-forming relation" in the next section.

**En-forming Relation**

As just noted, a material substance consists of a substantial form that en-forms matter. This brings us to what I call the *en-forming relation* between the form that en-forms and the matter that is en-formed. I use the phrase "en-forming relation" in reference to the relation the substantial form and matter of a material substance stand in. When the substantial form en-forms matter, the form and matter stand in an en-forming relation. This relation, on my view, is an explanatory relation that's *not* causal. Rather, this noncausal explanatory relation is a grounding relation.[9] The type of explanation grounding provides is distinct from scientific explanation or causal explanation; it provides metaphysical explanation (Fine 2012, p. 37). To elucidate this type of explanation, we'll consider two examples and then apply the concept of grounding to the en-forming relation.

First, imagine a halfpipe that is decorated with graffiti at the iconic Burnside Skatepark in Portland, Oregon. Suppose someone spray-painted a white, red, and black image of a bloodshot eyeball onto its surface, and thus the halfpipe is colored. One could explain why it's colored by referring to the graffiti artist's actions and the cans of spray-paint used to produce the effect of the skatepark feature being covered in white, red, and black paint. Such a causal explanation would be what most people are interested in.

But a noncausal metaphysical explanation could also be given for why the halfpipe is colored given that it is dressed in white, red, and black paint. The metaphysical explanation is that the halfpipe is white, red, and black, and white, red, and black are colors, and therefore the halfpipe is colored. Put differently, the halfpipe is colored in virtue of being white, red, and black. The reality that (*a*) the halfpipe is white, red, and black grounds (*b*) the reality that it's colored.

It is important to notice several things about the grounding of *b* in *a*. First, there's *asymmetry*. It is the case that *a* grounds *b*, but *b* does not ground *a*. After all, *b* does not explain *a*, since *b* could be true and *a* could be false. The halfpipe could have been painted different colors and *b* would be true in that case as well. Second, related to this asymmetry is *dependency*—*b* depends upon *a* but not vice versa. Third, notice that *a* is *explanatorily prior* to *b*. We can assume that *a* and *b* became actual simultaneously, and therefore *a* is not temporally prior to *b* (which we could assume if *a* provided the causal explanation of *b*). So the sense of priority that's relevant is not causal but rather explanatory. It is not that *a* is temporally prior to *b*; it's explanatorily prior. Given that *a* is true, *b* is true. The former explains the latter. Such noncausal explanatory priority is fundamental to grounding (cf. Correia and Schnieder 2012, p. 1).

Now, let's consider a second example involving David Beckham. Let us assume that he is, at present, a football player on the Manchester United football team. If we wanted to explain why Beckham is a Manchester United football player, we might give several different kinds of explanations. We could appeal to the lucrative salary he was offered to play for Manchester and the psychological influence that had on him, which led to his decision to sign with Manchester. We might call this an economic or psychological explanation. We also might give a physical explanation, such as he is wearing a Manchester jersey and he is passing the ball to other Manchester players.

Yet, a metaphysical explanation could also be given. Such as, Manchester United is a football team consisting of a set of twenty football players, and Beckham is a member of this set. In other words, Beckham is a Manchester United football player in virtue of being a member of a set that the Manchester United football team consists of. The reality that Beckham is a member of the set constituting the Manchester team grounds the reality that Beckham is a Manchester football player. Notice that this does not give us a causal explanation of why Beckham is a Manchester player. It gives us a metaphysical explanation in terms of grounding.

Suppose further that we wanted to know why Beckham is a professional football player. Once again, we could give a metaphysical explanation in terms of grounding. That is, Manchester United is a professional football team and Beckham is a Manchester United player, and therefore he is a professional football player. The reality that (*A*) Beckham is a member of the set of players that Manchester United consists of grounds (*B*) the reality that he is a Manchester player, which grounds (*C*) the reality that he is a professional football player. Here the fact that grounding can be transitive becomes relevant (see Correia and Schnieder 2012, p. 8). In this example, *A* grounds *B* and *B* grounds *C*, so *A* grounds *C*.

In sum, the type of explanation grounding provides is distinct from scientific and causal explanation. Grounding provides metaphysical explanation. Furthermore, grounding is asymmetric; it involves asymmetric dependence where what's grounded depends on what grounds it, and what grounds is explanatorily prior to what is grounded. And in some cases grounding is transitive.

There's much more that can be said about grounding, and there are issues that different proponents of grounding might disagree on. One issue, for example, is how grounding relates to modality. If *x* grounds *y*, is *y* necessary in every possible world that includes $x$? Or, is it possible for $x$ to exist without $y$ even though $x$ grounds $y$? Different proponents of grounding have different views about whether or not grounding includes necessity (cf. Audi 2012; Chudnoff 2011). Another topic of disagreement is whether or not grounding is analyzable in nongrounding terms or whether it is primitive and unanalyzable. Most proponents of grounding, but not all proponents, think grounding is primitive and unanalyzable (Bliss and Trogdon 2016, section 2; Correia and Schnieder 2012, p. 13). The claims I am defending in this work do not require any particular position on these issues. Therefore, I won't here defend any particular position on these issues.

Having introduced grounding, we can now return to en-forming. The en-forming relation between a substantial form and the matter it en-forms is a grounding relation. If substantial form $x$ stands in an en-forming relation to $y$, then $x$ grounds $y$. Since en-forming is grounding. The noncausal en-forming relation between a substantial form and matter is an instance of grounding. The form grounds the existence of the matter it en-forms.

As mentioned above, Aristotelian metaphysics includes the concept of prime matter, which is matter that is not en-formed, and therefore does not exist. Matter, on the other hand, does exist, and it exists in virtue of being en-formed. But matter never exists as undifferentiated matter. It exists as matter of a particular kind of material object. In the case of a material substance, the material entity that consists of the en-formed matter is a unified entity of a particular kind. In Aristotelian terms, the material entity is a unified entity with a particular *essence*, and this unified material entity exists in virtue of being en-formed by a substantial form.

The substantial form grounds the material entity's existence. And the substantial form grounds the material entity's existence by virtue of the substantial form grounding the unity and essence of the matter that the material entity consists of. So there's a transitivity of grounding: The substantial form grounds the unity and essence of the matter, *and therefore* it grounds the existence of the unified material entity of a particular kind.

According to the fundamental principle of neo-Thomistic hylomorphism, which will be discussed in much more detail below, the human soul en-forms the human body. The human soul stands in an en-forming relation to the matter it en-forms. The en-formed matter is the body. In light of our foregoing discussion, we can describe this en-forming relation the soul stands in to the body in terms of grounding. In short, the soul grounds the body. That is, the soul grounds the existence of the body. This is in virtue of the soul grounding the body's unity and essence.

For elucidation, consider the soul and body of a particular individual human. Allow "S" to stand for Socrates's soul and "B" to stand for his body, which is a unified entity that's a particular kind of thing, namely, a human body. "S *en-forms* B" means S *grounds the existence of* B; and S grounds the existence of B because S grounds B's unity and essence. The en-forming relation is a grounding relation.

As I move forward in describing and then applying neo-Thomistic hylomorphism, I will use the Aristotelian and Thomistic terminology introduced in this section, including "en-forming relation." Yet as I've discussed, what I mean by the claim that Socrates's soul en-forms his body is that Socrates's soul grounds the existence of his body. And as we will see, contra Plato and Descartes, the body is imperative on the view I'm advocating.

## HUMAN ONTOLOGY

This section will present an overview of the neo-Thomistic hylomorphic view of human ontology and clarify how it differs from various competing views. In the following section, I will focus more specifically on the nature of the soul on neo-Thomistic hylomorphism.

Some might find it odd that the study of human ontology and philosophy of mind will be done in tandem throughout this section and the next. Yet, the rationale for considering Aquinas's philosophy of mind and human ontology together is simple: that's how Aquinas himself approached the subjects. Although he may not have referred to them as "philosophy of mind" and "human ontology," he wrote a lot on these subjects. However, it does not seem that the two areas of study are disjointed independent areas of inquiry in Aquinas's mind. After all, his most elaborate discussions of the soul, its mental capacities, and how the soul relates to the body are found in a section of the *Summa Theologiae* known as the "Treatise on Human Nature" (see Aquinas 2002).[10] Hence, in this section, Aquinas's thoughts on human ontology and philosophy of mind will be discussed in tandem.

### Human Nature

It is well known, or at least often thought, that Descartes held that human persons are thinking substances, and thus identified human persons with their soul (cf. Descartes 1996, p. 54).[11] On this Cartesian view, it seems that there is a disconnect between the nature, or essence, of humanity and the body. Our body is in no sense essential to our humanity.

Such a notion is at odds with the intuition that our bodies are integral to us as human beings—an intuition that's evident when we make comments like "Jordan is two meters tall," "Aryn looks beautiful," or "Kai is paralyzed from the waist down." It is one's body that is a certain height, looks beautiful, or is paralyzed. Since it seems that a human person's body is an integral aspect of them, it's fitting to say a human person is a certain height, looks beautiful, or is paralyzed. The way that we talk about ourselves and our bodies suggests that it seems intuitive that our bodies are integral to us as humans. Cartesian dualism doesn't appear to fit well with this intuition.

Like Cartesian dualism, Platonic dualism also identified human persons with their soul (see Aquinas, ST 1a 75.4c; Stump 2003, pp. 191–192). This view was well known in the thirteenth century, and particularly well known, and *resisted*, by Aquinas (Stump 2003, pp. 192–194). The body, on Aquinas's view, is not an extra add-on to a human person like one's clothing. One respect in which hylomorphism differs from Platonic and Cartesian dualism is that the body is seen as essential to a human person.

Aquinas's most substantive teaching on human ontology is found in the aforementioned "Treatise on Human Nature" (cf. Pasnau 2012, p. 350). This work is in the first part (i.e., prima pars) of the *Summa Theologiae*, from questions seventy-five to eighty-nine. The *Summa* is a theological work meant to educate those who are just beginning theological study (see Pasnau 2002b, p. xiii). As Aquinas explicates his views throughout this work, he relies on three authorities: Aristotle, Church tradition, and Scripture.

As Eleonore Stump (2003, p. 192) points out, Aquinas's general thoughts on human nature were guided by intuitions summed up by two biblical passages: *Genesis* 3:19b and *Ecclesiastes* 12:7. The former reads, "for you are dust, and to dust you shall return."[12] God is here speaking to Adam as if Adam is dust, and thus a physical being. According to the latter verse, "and the dust returns to the earth as it was, and the spirit returns to God who gave it." These words are spoken by the Preacher, the son of King David about the death of man (see *Ecclesiastes* 12:8; 1:1). Interestingly, this latter passage suggests that human persons have a spiritual constituent that's not identical to the body and persists after death. Together these passages imply that human persons are physical beings, but not *merely* physical beings (see also Matthew 10:28; Philippians 1:21-24; 2 Corinthians 5:1-8). They can easily be taken to suggest Aquinas's view that a human person is composed of matter and soul.

After discussing purely spiritual creatures and corporeal creatures, Aquinas turns in the "Treatise on Human Nature" to consider human beings, "who are composed of a spiritual and corporeal nature" (Aquinas, ST 1a 75). The "spiritual nature" Aquinas is referring to corresponds to the soul. The "corporeal nature" corresponds to the body. He thought a human person is a material substance composed of matter that's the physical constituent of the body, and a soul that's the substantial form of the body. Reasoning from what he thinks is true of a particular human to what he thinks of humankind, Aquinas (ST 1a 75.4c) writes: "For just as it belongs to this human being to be composed of *this* soul, *this* flesh, and *these* bones, so it belongs to the account of human being to be composed of soul, flesh, and bones."[13] Regarding human nature, Aquinas agrees with Augustine who thought "a human being is neither the soul alone, nor the body alone, but the soul and the body together" (Augustine, *City of God*, XIX.3 in Aquinas, ST 1a 75.4sc).[14]

On another critical point mentioned above, Aquinas agrees with Aristotle, who thought the human soul must be the form of a material body (see Aristotle, DA 412a 20). As the form of the body, the soul "contains the body" and "makes it be one thing" (Aquinas, ST 1a 76.3c). Simply put, it unites the body by grounding the existence and essence of each of its parts. A human person, on this view, is a substance that naturally consists of a substantial form (i.e., the soul) that en-forms matter (i.e., the body). A human person consists of a soul that naturally en-forms its body, which is essential to human nature (cf. Pasnau 2012, pp. 349–350).

But why is the human soul naturally the form of a body that's essential to human nature? Couldn't it be nonessentially united to a body? Following Aristotle, Aquinas (ST 1a 76.1c) reasons:

> For the soul is the primary principle of our nourishment, sensation, and local movement; and likewise of our understanding. Therefore this principle by which we primarily understand, whether it be called the intellect or the intellectual soul, is the form of the body. This is the demonstration used by Aristotle (*De Anima* ii, 2).[15]

Here we are introduced to a fundamental reason Aquinas thought the body is essential to human nature and the human soul naturally en-forms a body. The reason is that the human soul depends on the body in order to operate consistently with its nature (Aquinas, ST 1a 84.4c & 89.1c; cf. ST 1a 75.7 ad 3).

An essential operation Aquinas often refers to is sensing. Humans, like animals, are sensory creatures (Aquinas, SCG II.57). According to Aquinas, human nature includes sense perception and the human soul by nature senses. Though the human soul is an intellectual soul with the capacity to reason, it is also a sensory soul with the capacity to sense. Aquinas did not think the soul could sense by itself without the body—rather it depends on the body to sense. Once again following Aristotle, he thought that in order to sense one must be moved by external objects, and therefore the soul needs something that can be so moved, which is the body (Aquinas, SCG II.57). For example, the heat from a campfire affects the body producing a sensation of warmth. Since the human soul is sensory and depends on the body to sense, human persons by nature consist of a soul that en-forms a body, as Aquinas (ST 1a 75.4c) explains:

> But any given thing is identified with what carries out the operations of that thing, and so a human being is identified with what carries out the operations

of a human being. We have shown, however, that sensing is not the operation of the soul alone [75.3]. Therefore since sensing is one of the operations of a human being (even if not one unique to humans), it is clear that a human being is not a soul alone, but something composed of a soul and body. Plato, however, since he claimed that sensing belongs to the soul alone, could claim that a human being is a soul using its body.

Here we see a dependence of the soul on the body for its operations, or the exercise of its powers. This idea of soul-body dependence will be taken up again in chapter 7, as it's integral to the explanation of neural correlates of consciousness that I present.

There is also a second line of rationale for the idea that the soul is by nature the form of the body, which is essential to human nature. It appeals to soul-body unity. If, as Platonists thought, the body is nonessential to human nature and the soul is not naturally the form of the body, then it is difficult to explain why the human soul is united to a body that's nonessential to human nature. As Stump (2003, p. 194) points out:

> According to Aquinas, Platonists will have trouble explaining the way in which soul and body are joined. Platonists, Aquinas says, are committed to supposing that the soul is united to the body through some intermediary, because diverse, distinct substances cannot be bound together unless something unites them.

Because the body is nonessential to human nature according to Platonists, it will be difficult for them to explain why the human soul is united to a body, and they will have to explain it by appealing to an intermediary between the soul and its body.

But if, as Aquinas held, the human soul is by nature the substantial form of a body essential to human nature, then it's only fitting for it to en-form a body and appealing to any intermediary that unites body and soul is unnecessary (see Aquinas, QDA 9c). Since Aquinas's view does not require postulating something that unites the human soul to its body, his view provides a simpler explanation of why the soul is united to a body. This issue of soul-body unity that Aquinas was concerned with is similar to the causal pairing problem, and, as will become apparent in chapter 6, my response to the causal pairing problem is similar to Aquinas's treatment of this issue. Yet for now, it's important to further explain hylomorphism before applying it to mental causation and NCC, which I will do in the next two chapters.

While human nature includes the body because humans are sense-perceptible and the soul depends on the body to sense, the body depends on the soul for its existence. The soul doesn't depend on the body for existence, but rather grounds the existence of the body. In Aquinas's (QDSC 2 ad. 3) words:

> It must be said that the soul has subsistent actual being, inasmuch as its own actual being does not depend on the body, seeing that it is something raised above corporeal matter. And yet it receives the body into a share in this actual being in such a way that there is one actual being of soul and of body, which is the actual being of a man.

According to Thomism a human person is one substance that consists of a soul that en-forms matter, which is the body that exists because it is en-formed by a soul. Given that the body's existence depends on being en-formed by the soul, it cannot subsist on its own apart from the soul that is its substantial form.

It's critical to see that on the hylomorphic view of a human being, the soul (i.e., the substantial form) and the body (i.e., the en-formed matter) are related via an en-forming relation. As the form of the body, the soul en-forms the body. The existence of a particular human body is grounded by a particular human soul. Were the body not en-formed, it would not exist. Not only does an individual human body have existence because it is en-formed by a particular human soul, it is also what it is because it's en-formed by a form of a certain type. Hillary Clinton's body, for example, exists since it is en-formed by a soul, it is a human body since it's en-formed by a human soul, and it is Clinton's body since it's en-formed by her soul. In short, Clinton's body exists and exists as her human body because her soul en-forms the matter her body is comprised of.

At this point, hylomorphists disagree with physicalists who think the existence of the mental ontologically depends on the physical. Such ontological dependence of the mental on the physical would safeguard the explanatory priority of physics, which is critical to physicalism (cf. Kim 1993, pp. 209–210). Thus, many physicalists who hold to mind-body supervenience see it as a relation of ontological dependence, where the mental is dependent on the physical (see Kim 2011, p. 12). For instance, "according to realizer functionalists, a mental property is to be identified with the first-order physical property occupying (or 'realizing') that mental property's defining causal role" (Tiehen 2012, p. 223). If that's the case, then the mental property could not exist without some physical base property realizing such a causal role. The mental, on this view, depends for its existence on the physical. Hence, the realizer functionalist sees the dependency relation going in the opposite direction than the Thomistic hylomorphist. For physicalists in general, mental states ontologically depend on the physical. For the Thomistic hylomorphist, the intellectual soul possessing mental capacities grounds the physical body's existence, and therefore the latter ontologically depends on the former.

Interestingly, hylomorphism also differs from dualist views that assign ontological priority to the physical, such as property dualism and emergent dualism. Property dualists admit sui generis mental properties, but see such properties as depending for their existence on the physical (cf. Chalmers 1996; O'Conner 1994). Similarly emergent substance dualists think a mental substance emerges from and thus ontologically depends on the physical (see Hasker 1999, pp. 189–190). Given that hylomorphism claims the dependency relation goes in the opposite direction because the soul grounds the body's existence, hylomorphism differs from property and emergent dualism.

Hylomorphism does not only distinguish itself from physicalist views or views with physicalist leanings. It also differs from Cartesian substance dualism. The body, on hylomorphism, is a substance only in the sense that it is an inseparable aspect of a single human substance.[16] Apart from being an aspect of a human substance, it is not a substance. In other words, it is not a substance in-and-of itself. However, according to Cartesian dualism, the body is a substance in itself, as the soul is.[17] Hence, you have two substances on the Cartesian view that exist in-and-of themselves. But according to hylomorphism, it's not that two substances, which exist in-and-of themselves, form one person consisting of two substances. Rather, a human person is one substance. The body is an inseparable aspect of that one substance, on neo-Thomistic hylomorphism.

Furthermore, the body is critically relevant to human nature from the Thomistic perspective. Accordingly, in *Disputed Questions on Spiritual Creatures* Aquinas (QDSC 2 ad. 5) writes:

> It must be said that no part has the perfection of a nature, when separated from the whole. And hence the soul, since it is a part of a human nature, does not have the perfection of its own nature, save in union with the body.

Since Aquinas sees the body as essential to human nature, it's fitting that a human soul devoid of its body is lacking as a human substance on his view. In reference to what is lacking, Aquinas employs the concept of the perfection of a substance's nature. The idea appears to be that a human soul lacks the perfection of its human nature when unembodied. According to Aquinas, the body is required for the perfection of human nature.

What Aquinas (QDSC 2 ad. 5) goes on to say in the same passage suggests why. He thought a thing is not "perfect in its own nature unless what is virtually contained in it can be actually brought out." And certain powers of the soul, Aquinas seemed to think, are "dependent on corporeal matter," for "powers which are acts of the organs flow from it." In other words, some actions of the body are the exercise of powers grounded in the soul. If such powers are essential to human nature, the soul needs the body to be perfected in nature. For without the body, the soul lacks its capacity to exercise such powers. Thus, elsewhere, Aquinas (ST 1a 76.4 ad 2) notes that the soul moves the body "through its potential for producing movement, the actualization of which *presupposes* a body."[18] Basically, a human person's body is needed to fully exemplify their human essence, which includes certain properties and powers.

By way of summary, according to this view, a human person is a hylomorphic object consisting of soul and body. Furthermore, the body is essential to a human person. This distinguishes hylomorphism from Cartesian dualism, but also prompts the questions: What qualifies hylomorphism as a substance dualist position? After all, hylomorphism shares common ground with materialist positions like animalism and Peter van Inwagen's view of human ontology.[19] Yet, my aim is to demonstrate that neo-Thomistic hylomorphism can overcome the chief contemporary concerns regarding *substance dualism*. While I gladly admit hylomorphists share common ground with materialists, in the following section I will clarify what makes neo-Thomistic hylomorphism a substance dualist position.

## INCARNATE SOULS

The nature of the soul on Aquinas's view is controversial, especially the issue of whether the soul is an immaterial substance. According to my interpretation of Aquinas described in this section, the soul is an immaterial, or non-physical, substance. Given that some will think I've misinterpreted Aquinas, I reemphasize the *neo*-Thomistic element of my neo-Thomistic hylomorphic view. The primary aim of this section is not to defend my interpretation, but rather to explicate the nature of the soul according to neo-Thomistic hylomorphism.

I will begin by considering the passage that will guide our exposition. In question 76.1 of the *Summa*, Aquinas addresses an objection that challenges his view that the soul is the form of the body on the grounds that the soul subsists, or is a subsistent (Aquinas, ST 1a 76.1 obj. 5). According to Aquinas (ST 1a 29.2c), a subsistent is a thing that has existence in itself, as opposed to having its existence in another.[20] The objection claims Aquinas's view that the soul is the body's form is incompatible with his view that the soul subsists (see Aquinas, ST 1a 75.2c). Aquinas (ST 1a 76.1 ad 5) responds as follows:

> (*Section I*) The soul shares with corporeal matter the existence in which it subsists: from that matter and from the intellective soul, one thing comes about.

This occurs in such a way that (*Section II*) the existence that belongs to the whole composite also belongs to the soul itself, something that does not occur in the case of other forms, which are not subsistent.

I have labeled two sections of Aquinas's response, which will guide our considerations respectively. We will begin with *Section I* and consider the soul as the form of the body. Then we'll move on to *Section II* and consider the soul in itself.

## The Form of the Body

I mentioned above that Aristotle and Aquinas considered the soul to be the form of the body. As such, the soul grounds the body's existence. We see this idea described in *Section I* of the passage above. In the description there is one thing that exists in itself—the soul. But also included in the description is something that depends on the soul for its existence. That is, the corporeal matter en-formed by the soul, which is the body.

The soul "shares" its existence with the body. The soul has existence that does not depend on the body, and the soul confers existence on the body. The soul grounds the existence of the body given that it grounds the body's unity making it a unified existing entity that's a human body as opposed to another kind of body. The soul exists in itself, whereas the body exists because the soul exists and grounds the unity and essence of the body. In that sense, the soul "shares" its existence with the body. Consequently, there is one hylomorphic object—a human person—that exists. The person consists of a soul that en-forms corporal matter, which is the body.

Like I discussed in the previous section, as the form of the body the soul is meant to be embodied, according to Aquinas (see ST 1a 76.2 ad 6). Since my soul is the form of my body, its natural state is to be en-forming my body. It is naturally related to my body via an en-forming relation, as form en-forming matter. According to hylomorphism, a soul is not related to a body through an external relation, such as a causal relation, that pairs an immaterial substance with a material substance. Rather, there is an en-forming relation where the soul en-forms matter, and therefore a single material substance exists.[21] This en-forming relation is an intrinsic relation because it's internal to one substance, rather than between two substances. As previously discussed, this en-forming relation is a grounding relation (see the section "En-forming Relation"). The soul grounds the body's existence.

The body, according to this view, would not exist as a single biological human organism were it not a unified entity. The form of the body, the soul, grounds the unity of the body. Moreover, there are other kinds of unified biological organisms. What grounds the fact that a body is a human body, as opposed to some other kind of biological organism, is the fact that a human soul en-forms it. Thus, the form of a human body—a human soul—grounds a human body's existence in virtue of grounding its unity and its essence.

Given that the substantial form grounds the unity of the body, one might think the form is like what contemporary metaphysicians call bare particulars or bare substratum. A key difference, however, is that bare particulars and bare substratum are *bare* and thus property-less (cf. Loux 2006, p. 84). Substantial forms are not bare. The form of the human body, the soul, has properties and grounds the essential properties and powers of the human being (see Pasnau 2012, p. 361). So the human soul of hylomorphism should not be thought of as a mere bare substratum, nor tantamount to a powerless free-floating shape (Pasnau 2012, pp. 352–353). Souls have properties and powers. Souls are agents with causal power (Pasnau 2012, pp. 352–353).

## The Soul Itself

Since we've fleshed out *Section I* of our guiding text, let us now consider *Section II*. Here we read, "The existence that belongs to the whole composite also belongs to the soul itself, something that does not occur in the case of other forms, which are not subsistent." According to Aquinas, the existence of the whole substance belongs to the soul itself. Assuming that, one naturally wonders: How can the existence of the whole belong to the soul without the soul being a substance? The answer reveals that Aquinas is at least an "Aristotelian dualist," as Pasnau puts it (2002b, p. xvii). In short, the soul is a substance.

In question 75.2 of the *Summa*, Aquinas addresses the question of whether the human soul is subsistent. When briefly stating his position in the said contra, he quotes Augustine providing an affirmative answer. Directly following, he concludes: "Therefore the nature of the human mind is not only nonbodily, but also a substance—that is, something subsistent" (Aquinas, ST 1a 75.2sc). Not everything that subsists is a substance, but if the soul is a substance, then the answer to the question of whether it subsists is affirmative. As Pasnau (2002a, p. 225) points out in his commentary on this text, Aquinas prefers to call the soul a subsistent and is likely bringing his terminology in line with Augustine's use of "substance." But Aquinas can do this without miscommunicating his view of the soul, because according to his view the soul is both a subsistent and a substance. As Pasnau (2002a, p. 225) explains:

> Strictly speaking, the two are not equivalent, and Aquinas always prefers to describe the human soul as subsistent. For something to be a substance, in his strict usage, it must not only subsist but also be the underlying subject of accidents (see I *Sent.* 23.1.1c, *QDP* 9.1c). . . . The soul *does* meet the second

condition, and so does count as a substance, but Aquinas regards *subsistence* as the more basic notion.

Indeed, the soul is a unique form. According to this reading of Aquinas, the form of a particular type of material substance—a human person—is itself a substance (cf. Pasnau 2012, p. 353).[22] Moreover, according to Aquinas, it is a rational (i.e., thinking) substance (see Aquinas, ST 1a 75.2c & 75.6c).[23] To boot, it's a subsistent that can exist apart from the matter it en-forms, that is, the body (see Brown 2005, p. 103). That the soul is capable of such is made clear in the sentence directly following *Section II* of our guiding text. Having just said the existence of the whole composite belongs to the subsistent soul, Aquinas (ST 1a 76.1 ad 5) writes: "And for this reason the human soul continues in its existence after the body is destroyed, whereas other forms do not." Aquinas thought the soul's natural state is to be embodied, or in other words, to be en-forming the body (see Aquinas, ST 1a 76.2 ad 6). Yet, he also thought it could exist unembodied between bodily death and being bodily resurrected.

To be clear, my aim is to defend a Thomistic view of the soul, *not* a view of the afterlife. I, too, think human souls persist after bodily death and before bodily resurrection (albeit not for the same reasons as Aquinas).[24] Nevertheless, I describe Aquinas's view of the afterlife, specifically what theologians call the intermediate state, in order to further clarifying the nature of the soul, on Aquinas's view. In order to embrace neo-Thomistic hylomorphism one need not embrace Aquinas's view of the afterlife.

Given the picture of the soul on Aquinas's view sketched thus far, and that mainstream contemporary philosophy favors physicalism, Aquinas is clearly heterodox. It appears that he has committed an unpardonable transgression by crossing the line into substance dualism. However, two superlative Aquinas scholars whose works I have here relied on—Robert Pasnau and Eleonore Stump—are reluctant to place Aquinas in the substance dualist camp.[25] Pasnau (2002b, p. xvii) admits that Aquinas is a dualist but makes it very clear that he is not the same type of substance dualist as Plato and Descartes (see also Pasnau 2002c, Ch. 2). Stump (2003, p. 212) is willing to admit, "Aquinas seems clearly in the dualist camp somewhere." Yet, according to Stump's (2003, p. 212) interpretation of Aquinas, the soul is subsistent but not a substance, in light of his criterion for what qualifies as a substance.

Nonetheless, Stump (2003, p. 212) admits that the soul can exist without the body on Aquinas's view, which suggests that it is more like substance dualism than property dualism. She then suggests:

> Maybe we should invent a new genus *subsistence dualism*, under which substance dualism will be one species and Aquinas's account of the soul another.

> But perhaps we need not be so fussy. It is clear that Aquinas's account of the soul is more nearly allied with substance dualism than with property dualism; and if we do not take "substance" in "substance dualism" too strictly (if it can include subsistent things that are not complete substances), then we can count Aquinas among the substance dualists. In that case, we ought to categorize Aquinas as a non-Cartesian substance dualist and put him in the camp of those opposed to physicalism. (Stump 2003, p. 212)

That's not a bad option. Given that the subsistent soul we are talking about seems very much like what contemporary philosophers of mind would call a substance, whether or not Aquinas would according to his technical scholastic terminology. However, after describing this option, Stump (2003, pp. 212–216) apparently rejects it and goes on to use Aquinas's view as an example which suggests the category lines between dualism and materialism/physicalism have been misdrawn.

I agree with Stump that we should question the current categorizations, and Aquinas's view certainly challenges the contemporary depiction of dualism. Given where the lines are currently drawn, however, and the description of substance dualism I provided in the first chapter, it seems undeniable that Aquinas is a substance dualist.[26] I find myself agreeing with the spirit of Edward Feser's (2009, pp. 162–163) following words:

> In their zeal to emphasize the differences between Aquinas's position and that of Plato and Descartes, some of his defenders have tended to insist that he was not only not a materialist, but not a dualist either. But this "pox on both houses" approach, motivated in part perhaps by a fear that contemporary philosophers might be too quick to dismiss Aquinas if he is labeled with the "D word," is not very plausible. As we've seen, Aquinas held both that the intellect is immaterial and that the soul survives the death of the body. Surely that counts as dualism by most people's reckoning, and certainly by the reckoning of most contemporary philosophers.... Better, then, just frankly to acknowledge the fact, and to defend Aquinas's position on its merits rather than pretend it is something it is not.

Defending Aquinas's unique dualist position, or at least my *neo*-version of it, on its own merits is what I intend to do. Before doing so, it will be helpful to succinctly summarize the position in the conclusion of this chapter.

## CONCLUSION

The aim of this chapter has been to introduce a unique substance dualist position that originated with Aquinas. Using his thoughts as a guide, I have laid out a position that I think is quite defensible, which we can summarize as follows:

> *Neo-Thomistic hylomorphism.* A human person is a single substance that naturally consists of a substantial form that en-forms matter. A human soul is a substantial form that en-forms matter constituting a human body. The soul and body stand in an en-forming relation. Thus a person's body is fundamentally related to her soul via an en-forming relation, which is a non-causal grounding relation. The body is a unified biological organism consisting of matter unified by its form, the soul. The soul is a nonphysical rational substance that's not identical to the body, which it en-forms. The soul has properties and powers and grounds the essential properties and powers of the person, including the body.

The following three chapters are devoted to demonstrating that neo-Thomistic hylomorphism can overcome substance dualism's paramount problems and resolve a practical concern. I will apply it to the causal pairing problem in chapter 6 and then neural correlates of consciousness in chapter 7 before finally focusing on the practical issue of empirically discerning and measuring consciousness in chapter 8.

## NOTES

1. Another common spelling is *hylemorphism*.
2. See Robert Pasnau (2002c, pp. 65–72), Alfred J. Freddoso (2012, p. 6, footnote 5), James Madden (2013, Ch. 8), and Peter King (2012).
3. See the following section "Forms" for an explanation of what an accident form is.
4. Quoted from the translation of the *Summa Theologiae* by the fathers of the English Dominican province (i.e., Aquinas 1947).
5. Italics mine. With the exception of this quote, all other references to *Questions on the Soul* (abbreviated QDA) are referring to James H. Robb's (1984) translation.
6. For an explication of grounding, see section "En-forming Relation."
7. "Natural" here is not meant to mean "physical." Rather, assuming the natural world includes metaphysical facts, the form is thought to provide a metaphysical explanation. So it would provide a natural metaphysical explanation in contrast to either a natural physical explanation on the one hand or a nonnatural, supernatural explanation (e.g., God sustaining it in existence) on the other hand. I will often use the term "natural" referring to the regular way things are given the nature of reality. So there can be natural physical facts, but also natural metaphysical facts.
8. Cf. Leftow (2010, p. 397), Feser (2009, p. 14), and Wuellner (1956, prime matter).

9. Here I am only claiming that the en-forming relation is a grounding relation, not that all grounding is relational. Proponents of grounding disagree about whether grounding *necessarily* involves a relation (see Bliss and Trogdon 2016, section 3).

10. Unless otherwise noted, quotations from the "Treatise on Human Nature" in the *Summa Theologiae* (questions 75–89) are from Robert Pasnau's translation (i.e., Aquinas 2002). Excerpts from Thomas Aquinas, *The Treatise on Human Nature*, translated by Robert Pasnau, Copyright © 2002 by Hackett Publishing Company, Inc. reprinted by permission of Hackett Publishing Company, Inc. All rights reserved.

11. My comments here pertain to the popular reading of Descartes and the view that is commonly labeled "Cartesian dualism." Yet, as Stump (2003, p. 512) points out, at times Descartes seems to think that a complete human person is a soul-body compound. It is possible that Descartes has been misunderstood. Nevertheless, historical merits aside, the "Cartesian" view I will be referring to throughout this work is the popular understanding of Cartesian dualism. For it is the popular version of Cartesian dualism that's almost always the focus when substance dualism is considered, and it's the popular version of Cartesian dualism that Jaegwon Kim (2005) focuses on when he presents the pairing problem.

12. Biblical quotations are from the English Standard Version.

13. Italics are not mine.

14. There are clearer passages in Augustine's works that Aquinas could have quoted. For example, in *The Trinity*, Augustine (1991b, p. 403) writes, "we could also define man like this and say, 'Man is a rational substance consisting of soul and body.' In this case there is no doubt that man has a soul which is not body and a body which is not soul."

15. Quoted from the translation of the *Summa Theologiae* by the Fathers of the English Dominican province (i.e., Aquinas 1947).

16. I frequently use the phrase "inseparable *aspect*" in reference to what metaphysicians call an "inseparable part," especially in the context of discussing substances. For the word "part" conjures up images of separable parts in the minds of most English speakers. And since an inseparable part is very different from a separable part, the word "aspect" is preferable when conveying the idea of an inseparable part.

17. At least this is how Cartesian dualism is often described.

18. Italics mine.

19. Regarding animalism, see Eric Olson (2007, Ch. 2). Regarding van Inwagen's materialist human ontology, see van Inwagen (1990; 1993; 2007).

20. For example, my dog Anselm is subsistent, but his tail is not subsistent. It exists only as a part of Anselm's body. Strictly speaking, his tail would not exist as his tail if it were separated from his body.

21. See section "Material Substance."

22. Howard Robinson (2020, section 1.2) seems to likewise understand Aquinas. For another view on whether the soul is a substance according to Aquinas, see Christopher M. Brown (2005, pp. 55–57).

23. As I'll discuss in chapter 7, according to Aquinas's view of cognition there's a role for the brain (see Aquinas, ST 1a 75.2 ad 3, 1a 89.1c). Nevertheless, he thought "the human being thinks, through the soul" (Aquinas ST 1a 75.2 ad 2).

24. I think such can be demonstrated on theological grounds, but not philosophical.

25. See also Alfred Freddoso (2012, p. 6, footnote 5) and James Madden (2013, Ch. 8).

26. See chapter 1, section "Defeating Dualism's Defeaters."

*Chapter 6*

# En-forming Causal Pairing

Now that the causal pairing problem and neo-Thomistic hylomorphism have been clearly presented, it is time to apply the latter to the former. The pairing problem is substance dualism's chief problem regarding mental causation (see chapters 3 and 4). At the root of the problem is the charge that if substance dualism were true, then mental causation would be incoherent since there is no way for immaterial, nonphysical minds to be causally paired with their effects. The purpose of this chapter is to show that neo-Thomistic hylomorphism is a substance dualist position immune to the pairing problem.

My aim is to show that hylomorphism[1] offers a coherent explanation of why a particular mind is causally paired with its effects and thus provides a solution to the pairing problem. This explanation (or account) of causal pairing capitalizes on the fundamental tenet of hylomorphism—the soul is the form of the body. I will present my hylomorphic solution to the pairing problem in the section "The Proposed Hylomorphic Solution." However, there are other alleged hylomorphic solutions to the pairing problem. Therefore, in the section "Reconsidering the Causal Pairing Problem," we will reconsider the pairing problem, and I will explain two other hylomorphic solutions to it in order to distinguish them from my own. At that point, my rationale for preferring my hylomorphic solution to the pairing problem over and above the other solutions will be given. Before wrapping up the section "Reconsidering the Causal Pairing Problem," I will also give three preliminary critiques of Kim's argument for the pairing problem that do not depend upon hylomorphism. After explicating my hylomorphic account of causal pairing in the section "The Proposed Hylomorphic Solution," I respond to the two strongest objections to my account in the section "Objections."

## Chapter 6

# RECONSIDERING THE CAUSAL PAIRING PROBLEM

As demonstrated in chapter 4, the causal pairing problem alleges that mental causation is metaphysically impossible for the immaterial minds of substance dualism.[2] The nature of causation and the nature of immaterial minds supposedly preclude the possibility of mental causation. In this section, I will reconsider Kim's argument for the pairing problem. First, a brief summary of the argument will be given before I discuss two hylomorphic responses that differ from my own response. Then I will clarify which premises of Kim's argument I will object to via my own hylomorphic response. Yet before I present my response, which aims at providing a solution to the pairing problem, this section will wrap up with preliminary critiques of Kim's argument. As previously mentioned in chapter 4, I focus on Kim's work on the causal pairing problem simply because I think his work is the best there is on the problem. In my opinion, when it comes to mental causation in general, Jaegwon Kim's works are the cream of the crop.

Let us summarize Kim's argument for the pairing problem (cf. Kim 2005; 2009). Of fundamental importance is the idea that causation requires a pairing relation between a cause and its effect. Unfortunately, for substance dualism there is no possible relation that could pair an immaterial, nonphysical mind with an effect, argues Kim. After all, spatial relations are the only type of relation that could possibly pair causes with effects. Yet immaterial minds are not spatial, and thus cannot stand in spatial relations. Since spatial relations are the only type of causal pairing relations, immaterial minds cannot stand in causal pairing relations. Therefore, Kim concludes, mental causation is metaphysically impossible for immaterial minds. I have formally reconstructed Kim's argument as follows[3]:

(1) For every case of causation there is a cause-effect pairing relation between cause and effect.
(2) All cause-effect pairing relations require spatial relations.
(3) An entity/event must be spatial to stand in any cause-effect pairing relation (from 2).
(4) Immaterial minds are not spatial.
(5) Immaterial minds cannot stand in any cause-effect pairing relation (from 3, 4).
(6) Therefore, mental causation is impossible for immaterial minds (from 1, 5).

Hylomorphists might respond to Kim's argument in various ways. It will be helpful to consider two noteworthy responses hylomorphists might give that differ from mine.

## Two Other Hylomorphic Responses

The objective of this subsection is twofold. For one, I hope to clarify two potential hylomorphic responses that differ from mine in order to distinguish my hylomorphic response. Additionally, I will point out weaknesses of these other two hylomorphic responses. I do think these deficiencies provide rationale for preferring my response. However, let me admit that such shortcomings fall short of rendering the other two responses dubious dead ends. My aim is not to show that the other responses are irredeemably hopeless. Rather, my twofold aim is to distinguish them from mine and to warrant serious consideration of my proposed hylomorphic response.

The first response we will consider exploits the hylomorphic tenet that a human person is a single material substance. According to this response, the pairing problem does not arise given that a human person is a single substance. The second response claims that the soul is spatial via the body, and therefore the soul meets Kim's prerequisite of spatiality for causality.

### First Response: Single Substance

As was stated in the previous chapter, according to neo-Thomistic hylomorphism a human person is a single material substance consisting of a form and matter. The soul, according to this view, is a substance in itself that en-forms matter, and thus there's one material substance. While the body is not identical to the soul, it is also not a distinct substance apart from the soul that en-forms it. Rather, there is one substance—a human person—and the body is an aspect of that one material substance. Pope Francis, for example, is a single human person consisting of a soul that en-forms matter, which is Pope Francis's body. Given this view of human persons, the hylomorphist can question the very grounds that prompt the pairing problem.

Recall from chapter 4 that Kim (2005, pp. 73–74) tees up his argument for the pairing problem by commenting on erstwhile arguments presented by historical critics of Descartes's so-called "interactionist dualism." According to common renditions of this dualist view, the immaterial mind and material body causally interact with one another (Kim 2005, p. 77). Hence the infamous objection: The two substances—that is, the immaterial mind and the physical body—are of such different categories that it seems impossible for them to causally interact (see Kenny 1968, pp. 222–223). As an argument against dualism this objection "is incomplete and unsatisfying . . . it only expresses a vague, inchoate dissatisfaction of the sort that ought to prompt us to look for a real argument," writes Kim (2005, p. 74). For the sake of making progress toward such an objective, Kim (2005, p. 74) asks: "Why is it incoherent to think that there can be causal interaction between *things* in 'diverse categories'? Why is it 'impossible' for *things* with diverse natures to enter into causal relations with one another?"[4] Notice the use of the plural "things" and the concern regarding how these things interact. This is Kim's starting point.

The hylomorphist, however, is not obliged to start in the same place. Given the hylomorphic view that a human person is a single material substance, the hylomorphist need not start with *things*, plural. She begins with *thing*, singular. And since there is a single thing, not thing*s*, you do not have causal interaction between two things, according to hylomorphism (see Feser 2009, pp. 166–167; Madden 2013, p. 275). Instead, you have an exercise of power by one substance with distinct aspects.

For example, imagine that my soul manifests a mental intention to run that leads to my legs striding forward. The mental intention to run and the movement of my legs are the intention and movement of the same material substance exercising its power to run. There are not two substances causally interacting, according to hylomorphism. Instead, there is one substance manifesting its power to run that involves the mental intention and the movement of the legs.

Hylomorphism differs from Cartesian dualism in a significant respect. Cartesian dualism says two substances—soul and body—causally interact. But according to hylomorphism, there is no such interaction. And given that there is not interaction between two substances, there is no reason to ask what causally pairs the two interacting substances. At least, that's the central claim of this first response.

Put differently, the starting point of Kim's argument for the pairing problem is available given Cartesian dualism, but unavailable given hylomorphism. This is problematic for Kim's argument because he is ultimately trying to show that if substance dualism were true, then mental causation would be incoherent. To do so, he starts with tenets of a dualist view and then tries to show how mental causation would be metaphysically impossible given such tenets. Kim's rendition of Cartesian dualism supplies him with the tenets he needs for his starting point. For according to Cartesian dualism, there are two substances and they causally interact. Hylomorphism, however, does not supply Kim with such tenets. Because, according to hylomorphism, there is only one substance exercising its powers; there is no causal interaction between two substances. Nevertheless, Kim needs his starting point, and therefore he needs the tenets that provide it. Since hylomorphism doesn't provide such tenets, the pairing problem simply doesn't arise for the hylomorphist. Hence, it seems as though there is no pairing problem for hylomorphism.

That is the first hylomorphic response. It clearly hinges on the hylomorphic tenet that a human person is one material substance. Given that tenet, Kim's starting point for generating the pairing problem is absent and the pairing problem does not arise. If the hylomorphist intends to show that her position is consistent given her starting assumptions, which significantly differ from the tenets Kim relies on, then this response is worthy of serious consideration. However, this first response may be less persuasive to physicalists than my own since it highlights the different starting points, rather than highlighting common ground and then proceeding.

When it comes to persuasive power, the hylomorphist might be better off if she can grant some ground to Kim and still provide a satisfactory response. One aim of my own response, though not the primary aim, is to meet this desideratum. The hylomorphist that adopts my response can easily grant the validity of Kim's argument in addition to his first and fourth premises, and then merely take exception with his second premise, which entails his third. In light of this, my hylomorphic response to the pairing problem has the potential to be more persuasive than this first response. In due course, I will present my response and the reader can judge for herself. But beforehand we shall distinguish a second hylomorphic response, which also differs from my own.

*Second Response: Spatial Soul*

If the hylomorphist *primarily* wants to meet Kim on his own terms, one tempting response is to acknowledge the validity of his argument and accept all but one premise—premise 4. According to this premise, immaterial minds are not spatial. The hylomorphist that takes this route of rebuttal denies premise 4 since she thinks the soul is spatial via the body.

Yet, how can the hylomorphist say the soul is spatial via the body? Basically, she tries to capitalize on the idea that the body, which is spatial, is an aspect of the soul.[5] Given that the body is spatial and an aspect of the soul, she concludes that the soul is spatial since an aspect of it (i.e., the body) is spatial.

Recall that according to neo-Thomistic hylomorphism, the soul is a substance in itself and it's also the form of the body. The human soul en-forms matter that is the human body.[6] Thus, a human person is a particular type of substance, a material substance. The soul en-forming the body is a material substance since the substance includes en-formed matter. Yet it is important to note that the matter is not a substance in itself apart from the form that en-forms it, that is, the soul. After all, there is only one substance, and that is the soul. The en-formed matter, the body, is an aspect of the soul. Since the body is an aspect of the soul and the body is spatial, the soul is spatial since an aspect of it is spatial. In other words, the soul is spatial via the body. If that's true—that is, if the soul is spatial—premise 4 of Kim's argument is false. Consequently, premise 5 does not follow, nor does the conclusion. In brief, the pairing problem says: spatiality is necessary to stand in causal pairing relations, and since immaterial minds are not spatial, they cannot stand in such relations. If the hylomorphist can justifiably deny Kim's claim in premise 4 that immaterial minds are not spatial, it seems that she can undermine Kim's argument for the pairing problem.

As a hylomorphist, and as someone who is spatially located, I see the initial attraction of this route of response. For one, it seems obvious to me that I'm spatially located in Washington State at present, not in Antarctica. Secondly, it's perfectly consistent with hylomorphism, and Aristotelian thought at large, to think our souls are spatially located where our bodies are. After all, according to hylomorphism, a person's soul unites the parts of her body making it a unified object, so it is fitting to think her soul is spatially located where her body is.

Nevertheless, I doubt this second response can be successful. Suppose Kip's hand rises as the result of his mental intention to raise his hand and flag a taxi. This second response says that Kip's intention is causally paired with his hand rising because his soul is spatial, and thus his soul's intention stands in a spatial relation to the effect. However, the soul is spatial via the body, in the way described above. So, according to this response, Kip's soul is causally paired with the effect of his hand rising because his soul is spatial since his body, an aspect of Kip, is spatial. On this line of thought, it seems that some spatial part of the body stands in a spatial relation with the effect and that spatial relation explains the causal pairing relation. If that is the case, ultimately the spatiality of the body is what explains the bodily effect (i.e., Kip's hand rising). And we have just causally paired one spatial part of the body with another spatial part of the body.

Given that, we must ask: What causal work is the soul doing? The answer seems to be: none. After all, the soul is spatial since an aspect of the human substance—that is, the body—is spatial. Simplistically put, the soul's spatiality is the body's spatiality. Thus, it must be some spatial part of Kip's body that stands in the spatial relation to the effect, and therefore stands in the causal pairing relation. But if it is a spatial part of Kip's body that stands on the cause side of the cause-effect pairing relation, then a part of Kip's body is doing the causal work, not Kip's soul. So what causal work is the soul doing? It seems it is not doing any causal work, given the premises of this response.[7] But this response is supposed to explain how the soul, not a part of the body, is causally paired with effects.

At the end of the day, I think this second response is unpromising because a part of the body, rather than the soul, ends up being the cause. Whereas the first response lacked persuasive power since it focuses on the different starting points, leaving the hylomorphist with little common ground upon which to reason with her interlocutor. Given the deficiencies of these two hylomorphic responses, I offer a different hylomorphic response to the pairing problem. In the section "The Proposed Hylomorphic Solution," my hylomorphic response will be given. Before presenting it, however, I will offer preliminary reasons to doubt Kim's second premise.

## Questioning the Necessity of Spatial Relations

My hylomorphic response to the pairing problem will question Kim's second premise, which says all cause-effect pairing relations require spatial relations. This premise is essential to Kim's argument and entails premise 3, which says that an entity/event must be spatial in order to stand in any cause-effect pairing relation. Essentially, I will try to show that, contra Kim's second and third premises, hylomorphism can account for why a mental cause is paired with its effect without appealing to spatial relations. Yet before giving an explicitly hylomorphic response to the problem, it will be helpful to weigh some general reasons to doubt that spatial relations are necessary to pair causes with effects. In this subsection, such general reasons, which do not depend on hylomorphism, will be given.

To begin with, Kim's rationale for premises two and three will be clarified. Then I will point out an unstated assumption that Kim's argumentation rests upon. Subsequently two justified beliefs that are inconsistent with the idea that spatiality is necessary for causality will be brought to bear on the issue, and I will unearth a threat posed to premises two and three by Kim's own view of qualia. Lastly, a flaw regarding the logical structure of Kim's reasoning will be pointed out.

### Kim's Case for Premises 2 and 3

Let's consider Kim's justification for his second major premise, which entails his third major premise. The following represents Kim's rationale for premise 2, which says all cause-effect pairing relations require spatial relations:

2.1 In clear cases of physical causation, spatial relations account for cause-effect pairing relations.
2.2 Apart from spatial relations, we do not know what else could account for cause-effect pairing relations.
2.3 Therefore, spatial relations are necessary for all cause-effect pairing relations.

Notice that 2.3 does not necessarily follow from 2.1 and 2.2, so we should not take this as deductive reasoning. It is either inductive or abductive reasoning for the idea that spatial relations are required for cause-effect pairing relations.

Secondly, let's note that Kim gives justification for each of these sub-premises. In support of 2.1, Kim (2005, pp. 78–79) offers an example of physical causation involving two guns, A and B. The guns are simultaneously fired, and simultaneously cause the deaths of Adam and Bob. Spatial relations, such as the distance gun A is from Adam and its orientation toward Adam, explain why the firing of A caused Adam's death. The same can be said for the firing of gun B causing Bob's death. Therefore Kim (2005, p. 79) concludes, "spatial relations seem to serve as the 'pairing relations' in this case, and perhaps for all cases of *physical* causation involving distinct objects."[8]

Perhaps, spatial relations do serve as pairing relations for all cases of *physical* causation. Yet, this hardly supports the idea that spatial relations must serve as pairing relations for all cases of causation, including those involving immaterial causes. Hence the need for 2.2. In support of this sub-premise, Kim (2009, pp. 32–33) offers the following line of reasoning:

> I can state my fundamental assumption in general terms, and it is this: It is metaphysically possible for there to be two distinct *physical* objects, $a$ and $b$, with ... the same causal potential or powers; further, one of these, say $a$, causes a third object, $c$. ... Now, the fact that $a$, but not $b$, causes $c$ to change must be grounded in some fact about $a$, $b$, and $c$. Since $a$ and $b$ have the same intrinsic properties, it must be their relational properties with respect to $c$ that provide an explanation of their different causal roles vis-à-vis $c$. What relational properties, or relations, can do this job? The *only* plausible answer seems to be that it is the spatial relation between $a$ and $c$, and that between $b$ and $c$, that are responsible for the causal difference between $a$ and $b$ vis-à-vis $c$ ($a$ was in the right spatial relation to $c$; $b$ was "too far away" from c to exert any influence). At least, there is *no other possible explanation* that comes to mind.[9]

The issue in this scenario is: Why did one *physical* object, rather than another *physical* object with the same causal powers, cause an event? Kim wants an explanation of what grounds the fact that $a$, but not $b$, caused a particular event. He thinks this explanation must depend on a particular type of relation that the physical objects stand in. The only answer that comes to his mind relies on spatial relations.

However, if the question included immaterial souls (rather than physical objects) as the alleged causes, then another relevant relation might come to mind for those sympathetic to dualism. Perhaps, an immaterial mind is paired with bodily effects simply in virtue of the relation of unity between a person's mind and body. For example, Smith's mental intention could be causally paired with Smith's hand rising because Smith's soul is united to his body. Likewise, Jones's intention might be paired with his hand rising since his mind is united to his body.

But the problem is that one's soul and body are united via a causal relation, according to Kim's reading of Descartes. So "the 'union' of a mind and a body that Descartes speaks of, therefore, presupposes mental causation" (Kim 2005, p. 77). Given this, Cartesian dualists would in effect be saying that Smith's mind causes his body because his mind causes his body. This circular explanation won't do. And supposedly there are no other relations that are good candidates for pairing causes with effects. Yet in causal cases involving physical objects, spatial relations seem to pair causes and effects. So spatial relations are the only relations that could pair causes with effects, even when it comes to mental causation involving immaterial minds. At least that's Kim's rationale for premises two and three, which claim spatial relations are required for causal pairing relations, and hence only spatial entities can stand in causal pairing relations.

*Unstated Assumption*

Kim (2005, p. 87) ultimately concludes that if substance dualism were true, mental causation would be precluded by the "diverse natures" of mind and matter. Kim thinks that the natures of immaterial minds, if such exist, must be significantly different than the nature of matter. Given the diverse natures of physical substances versus nonphysical substances, it seems fitting to assume that the nature of causation involving a physical cause is likely different than that of causation involving a nonphysical cause. However, Kim assumes the opposite.

He assumes that the principles governing physical causation must also govern causation involving nonphysical causes. From the outset he presupposes that examples of physical causation are paradigmatic for all causation, even causation involving immaterial substances. When Kim (2009, p. 32) attempts to prime the pump of our intuitions, he asks us to "consider an example of physical causation." Although, why should we expect that causation between merely physical objects would provide us with a causal framework that applies to causation involving an immaterial substance? After all, if immaterial substances are so different in nature from material substances, shouldn't we expect the nature of causation involving immaterial substances to be different than the nature of causation involving merely physical objects? We should at least be open to the possibility.

Yet, Kim tacitly assumes that causation produced by nonphysical substances must resemble causation produced by physical substances. He assumes that causation involving immaterial minds will be based on the same principles as causation between merely physical objects. So after using his examples of physical causes to show us that spatiality is necessary in such cases, he concludes that spatiality is necessary in every case of causation. Without Kim's unspoken assumption, the steps he takes to arrive at his conclusion are too hasty and unjustified. For causation between merely physical objects could be exactly as Kim describes it, and yet causation involving nonphysical substances could be of a different nature. Put differently, Kim may have identified the essential principles and necessary conditions for physical causation, but that does not justify the idea that causation involving immaterial minds must be based on the same principles and require the same conditions. Kim just assumes the latter.

At the very least, Kim owes us justification for his unstated assumption, which doesn't fit well with the idea that physical and nonphysical substances are very different in nature. Such justification may be hard to find apart from presupposing physicalism. Nevertheless, it's needed.

*Inconsistent Beliefs*

Some might think Kim's claim that an entity must be spatial to stand in a causal pairing relation does not demand much justification because it is obvious. Indeed, for the materialist, it may seem obviously true. After all, there may be nothing else in her ontology that could possibly account for causation. However, the idea that spatiality is necessary for causality seems much less plausible when it's shown to be inconsistent with justified beliefs. I will now point out two justified beliefs that seem to be inconsistent with the idea that entities must be spatial to stand in causal pairing relations. The first belief is classical theism (cf. chapter 3, section "Reasons to Think Closure is False"). The second is Kim's own belief about qualitative states of consciousness, coupled with the fact that pinpricks cause pain. I won't here defend the justification of these beliefs.[10] I will merely assume they are justified. My aim is to point out that, given these beliefs, Kim's second and third premises are not obviously true, but rather seem false.

To begin with, classical theism contradicts Kim's claim about the necessity of spatiality. According to classical theism, God is nonspatial and the creator of the universe. As such, God is the cause of the universe and all of spacetime coming into existence.[11] Thus, according to theism, God is the nonspatial cause of the universe and spacetime. This clearly contradicts Kim's claim that an entity must be spatial to be causally paired with an effect. Therefore, assuming that the theist's belief is warranted or justified, she can justifiably deny Kim's second and third major premises.[12] After all, she is justified in believing that there was a nonspatial cause of the universe and spacetime, and therefore she is justified in believing that a nonspatial entity can stand in a causal pairing relation. Again, I have not argued here that theistic belief is justified. Rather, the point is: assuming theistic belief is justified, theists can justifiably deny Kim's second and third premises.

Secondly, Kim's own view of qualia seems inconsistent with his claim that spatiality is necessary for an entity/event to stand in a causal pairing relation. At the end of *Physicalism, Or Something Near Enough*, Kim (2005, p. 170) concludes that qualia are physically irreducible. As discussed earlier, qualia are the felt sensations of what-it's-like to be in particular mental states. For example, a state of pain caused by a pinprick feels different than tasting chocolate. The difference in feeling is a difference in qualia. The point that concerns us here is that Kim thinks qualia are physically irreducible. In other words, according to Kim, qualia are nonphysical mental states.

Recall that Kim thinks nonphysical minds would be nonspatial if they existed, given that they would be nonphysical. Additionally, he does not think that there is a way of locating such nonphysical minds in space that would help dualists overcome the pairing problem (see Kim 2005, pp. 88–90). Assuming that Kim would consistently apply his thinking about nonphysical minds to nonphysical qualia, it's fair to suppose that nonphysical qualia would be nonspatial according Kim's view. Granted, he does not explicitly say this, but if there is an advantageous way for him to locate nonphysical qualia in space, then it seems that dualists could do the same regarding nonphysical minds. Yet, according to Kim, dualists cannot do so; thus, it seems that nonphysical qualia cannot be spatially located, according to Kim's line of thought. One might object that if qualia are properties rather than substances, then they can be spatially located where their bearers are. However, there are at least two considerable responses to this objection.

The first response is that Kim's commitments apparently exclude this solution. Kim claims that if souls were spatial, then spatial exclusion would apply to them and they couldn't be co-located in the same space as their bodies (see Kim 2005, pp. 89–90). He seems to think spatial exclusion applies to anything and everything that is spatial. Assuming such a position, if qualia are spatial, then spatial exclusion applies to them in addition to their physical bearers. As a result, qualia could not be co-located where their physical bearers are any more than spatial souls could be co-located where their bodies are. Thus, given Kim's commitments, the prospect of co-locating qualia where their physical bearers are faces the same issue that Kim alleges Cartesian souls face. Granted, this is not necessarily fatal for nonphysical qualia or souls. Perhaps nonphysical qualia could be spatially located in relevant proximity to their physical bearers, if not in the exact location. And if mental causation does depend on such spatiality, then the necessary condition would be preserved. However, if such is possible for nonphysical qualia with respect to their bearers, I see no reason why it could not be possible for nonphysical souls. In other words, if nonphysical qualia can be spatially located in some helpful way to preserve mental causation, then the same is probably true of souls.

This brings me to my second response. Kim makes no attempt to preserve the causal efficacy of qualia, which is what would motivate an attempt to locate qualia in space. In fact, Kim (2005, pp. 170–171) cuts the causal cord and surrenders nonphysical qualia, the irreducible "mental residue," to epiphenomenalism. According to epiphenomenalism, there are sui generis mental properties that are causally impotent. Regarding nonphysical qualia, Kim clarifies: "It has no place in the causal structure of the world and no role in its evolution and development" (Kim 2005, p. 171).

Kim's willingness to give up on the causal efficacy of qualia is consistent with his aims. After all, if you allow nonphysical qualia to have causal jobs, you're cracking the door open for other nonphysical aspects of reality to obtain causal roles. To boot, that might take causal work away from physical entities. "May it never be!" says the primacy of physics. Put differently, if nonphysical qualia can be causally efficacious that compromises the causal closure of the physical domain, which would compromise the primacy of physics (cf. Kim 1993, pp. 209–210). By claiming that qualia are epiphenomenal, Kim avoids this possibility. Moreover, assuming that qualia are epiphenomenal, he has no need to explain how they are spatial and thus capable of causation. In sum, Kim's tenets imply that nonphysical qualia are nonspatial and he has no reason to locate them in space, but he has reason not to.

Kim's tenets have not been haphazardly adopted, and they are needed to secure the causal pairing problem. However, Kim's commitments imply that nonphysical qualia are nonspatial, and here's the rub: Qualia often stand in cause-effect pairing relations. Pinpricks often cause pain, to give just one example. The felt sensation of what-it's-like to be in such pain is a qualitative state, a quale.[13] Thus some qualia—that is, states of pain—are caused by pinpricks. So nonphysical qualia, which Kim's tenets suggest are nonspatial, stand in cause-effect pairing relations, albeit on the "effect" side of the relation. But according to Kim's second and third major premises, nothing can stand in a causal pairing relation if it is not spatial. Therefore, if Kim is correct about the irreducibility of qualia, then qualitative states of pain caused by pinpricks refute his second and third major premises.

To conclude, Kim's own view of qualia coupled with the fact that pinpricks cause pain provides reason to doubt Kim's second and third major premises. For there is a clear inconsistency between these premises and the nonspatiality of irreducible qualia caused by pinpricks. Hence, it is not only that these premises seem false from the theist's vantage point, Kim's own view of qualia justifies doubt regarding these premises. Granted, those who don't share Kim's view of qualia will be unmoved by this point. However, the idea that qualia are irreducible is not without warrant and it is gaining noteworthy traction in philosophy of mind (see Chalmers 1996; 2003; Kim 2011, Ch. 10; Robinson 2012).

*Non Sequitur*

In addition to the aforementioned issues, Kim's second major premise does not actually follow from the justification he offers for it. His first sub-premise, 2.1, says that we know that spatial relations pair physical causes with physical effects. His second sub-premise, 2.2, says that we do not know what other type of relations could pair causes with effects. This suggests the following: If there are nonphysical causes, we do not know what type of relation would pair such causes with their effects. That's inconsequential for the prospect of nonphysical causes. After all, there are many facts about the world that we are ignorant of, and one such fact could be what type of relation pairs nonphysical causes with effects.

Of course, Kim thinks that what follows from 2.1 and 2.2 is that all pairing relations require spatial relations. But that does not actually follow. That would only follow if we necessarily know what type of relation would pair all causes with their effects. Yet it is possible for there to be a cause-effect pairing relation that we are presently ignorant of. Kim's conclusion only follows if we know, in every case of causation, what relation pairs the cause with its effect. However, we may not have such knowledge with respect to some cases of causation. Such lack of knowledge on our part would do nothing to affect the existence of causal pairing relations that we are yet to understand.[14] All that would follow is that there is more for us to learn.

Furthermore, it may be that when the physicalist considers what type of relation could possibly pair a nonphysical cause with its effect, no other type of relation other than a spatial relation "comes to mind," as Kim (2009, p. 33) put it. However, for this hylomorphist another type of relation does come to mind—that is, an en-forming relation. The en-forming relation that the soul stands in to the body, according to neo-Thomistic hylomorphism, provides the basis of my hylomorphic solution to the pairing problem.

## THE PROPOSED HYLOMORPHIC SOLUTION

Above I distinguished two hylomorphic responses to the pairing problem that differ from my own hylomorphic response (see section "Two Other Hylomorphic Responses"). I also gave reasons, which do not depend upon hylomorphism, to doubt Kim's rationale for his second major premise that says spatial relations are required for causes to be paired with effects (see the section "Questioning the Necessity of Spatial Relations"). Now in this section I will give my own unique hylomorphic response to the pairing problem.[15]

My response denies premise 2, which entails premise 3, of Kim's argument and justifies such a denial by showing that the hylomorphist can appeal to an en-forming relation to account for causal pairing.

In a nutshell, my proposed solution says that a person's soul is causally paired with her body in virtue of the en-forming relation that her soul stands in to her body.[16] In other words, the mental intention (e.g., to raise one's right arm) of a person's soul is causally paired with physical effects in her body (e.g., her right arm rising) since her soul en-forms her body. One's soul is causally paired with her body not because of a spatial relation, but rather as a result of her soul en-forming her body. Given this alternative account of causal pairing, the hylomorphist can justifiably deny Kim's second major premise, which entails his third, while agreeing with Kim's first and fourth major premises.

This section will be entirely devoted to presenting and explaining the proposed hylomorphic solution. Following this section, I will explicate and respond to the two most forceful objections to my hylomorphic account of causal pairing.

### En-forming Relation and Causal Pairing

The causal pairing problem allegedly shows that if substance dualism is true, mental causation is metaphysically impossible. This impossibility is said to arise because nonphysical minds, or souls, are nonspatial. Such nonspatiality is critical because, according to Kim, spatial relations are the only relations that could possibly pair mental causes with effects. Thus, nonphysical, nonspatial minds cannot stand in causal pairing relations. They are disqualified from all causal work and specifically mental causation. That's the problem. And it hinges on Kim's second premise—all cause-effect pairing relations require spatial relations—which is true if, and only if, spatial relations are the *only* relations that can pair causes with effects.

The Cartesian dualist might be inclined to think that a causal pairing relation can be explained by appealing to a relation of mind-body unity. However, Kim makes a significant point that is worth recalling once again. On his reading of Descartes, Cartesian dualism says a person's mind is united to their body via a causal relation (Kim 2005, p. 77). In other words, a mind-body relation of unity is explained by a mind-body causal pairing relation. Given that, if Cartesians try to explain a causal pairing relation between one's mind and body by appealing to mind-body unity, they will rely on the very pairing relation they're attempting to explain.[17] As Kim (2005, p. 77) points out, such an explanation "presupposes mental causation." Consequently, Cartesians cannot appeal to mind-body unity to explain causal pairing. For mind-body unity itself is supposedly explained by causal pairing. Since the causal pairing relation is explanatorily prior to the relation of mind-body unity, the latter can't be appealed to in order to explain the former. Hence the problem for Cartesian dualism, at least as Kim depicts it.

On the neo-Thomistic hylomorphic view introduced in the previous chapter, there is a very different account of mind-body unity. According to Aquinas (ST 1a 76.3c), Plato thought the soul is united to the body as mover to moved. This Platonic view is much like Kim's description of the Cartesian account of mind-body unity. Aquinas (ST 1a 76.1c, 76.6c, 76.7c) explicitly rejected Plato's account, and he postulated that the soul is "immediately united to its body as form to matter" (Aquinas, ST 1a 76.7c).[18] Simply put, the soul is united to its body as its form (Aquinas, ST 1a 76.6 ad 3).

The mind, or soul, is *not* united to the body via a causal relation. Rather, the body is en-formed matter, and the form that en-forms that matter is the soul. Therefore the soul stands in an en-forming relation to the matter of the body. The soul is the form of the body. As such, the soul grounds the existence of the body as a unified biological organism of a particular kind (i.e., a human body). The soul, qua form, grounds the unity and essence of the body. Without the soul en-forming the body, the body would not exist.

Let's use Kim's hypothetical characters, Smith and Jones, to illustrate the form matter en-forming relation as it pertains to human persons. According to neo-Thomistic hylomorphism, Smith's soul en-forms his body. This means that Smith's body is a unified entity and has the essential properties and powers it has due to the form that en-forms it—that is, Smith's human soul. The existence of Smith's body depends on Smith's soul en-forming it. If the matter of Smith's body wasn't en-formed by Smith's human soul, his human body wouldn't exist. Likewise, Jones's body exists as a unified whole that's his human body since his soul en-forms it. The existence of each person's body is grounded by the human soul that en-forms it. Consequently, each person's soul is united to their body as the form of their body.

The critical point for our purposes is that each human soul stands in an en-forming relation to the matter of the body it en-forms. According to this hylomorphic view, the en-forming relation is the most fundamental relation the soul stands in to the body. Moreover, the soul is the form of the body before it is the cause of bodily movements. But let me be clear that I'm not speaking here of temporal priority, since that's not my concern. The en-forming relation is explanatorily prior to any causal relation the soul stands in to the body that it en-forms. Given that, we might deduce that no causal relation would be temporally prior to the en-forming relation either. Nonetheless, what's imperative is that the soul stands in an en-forming relation to the body *explanatorily* prior to any causal relation.

In light of this, let us reconsider Kim's (2005, p. 76) thought experiment involving Smith and Jones. Recall that these two unfortunate individuals are psychologically synchronized. Whenever Smith wills to raise his hand, Jones wills to raise his hand as well. As a result, their hands rise simultaneously. This prompts Kim's (2005, p. 76) vital question: "So why is it not the case that Smith's volition causes Jones's hand to go up, and that Jones's volition causes Smith's hand to go up?" Put differently, why is Smith's soul, and thus his volition, causally paired with the rising of his hand, not Jones's?

The hylomorphist has an answer: Smith's soul en-forms Smith's body, not Jones's body, and therefore Smith's soul is causally paired with his body, not Jones's. On this view, the fact that Smith's soul en-forms Smith's body explains why his volition is causally paired with his hand going up. Basically, his soul is causally paired with his body since his soul en-forms his body. On the other hand, Jones's soul en-forms Jones's body, and therefore his soul is causally paired with his body.

It is perfectly fitting that a particular body that's en-formed by a particular soul would be causally paired with the soul that en-forms it. It would be extremely odd if Donald Trump's soul en-formed his body but was causally paired with Theresa May's body that's en-formed by May's soul. To the contrary, if Trump's soul en-forms Trump's body, we would expect his soul to be causally paired with his body. Hence, this hylomorphic explanation easily avoids being ad hoc. The hylomorphist is explaining causal pairing by appealing to the en-forming relation that is absolutely fundamental to Aristotelian-Thomistic human ontology.

Another benefit of the hylomorphic account I'm offering is that it's not circular. It does not presuppose a causal pairing relation that allegedly accounts for mind-body unity in an attempt to explain a causal pairing relation, as Cartesians are accused of doing. For the en-forming relation that the soul stands in to the body is explanatorily prior to causal pairing relations. In the explanatory chain, Smith's soul is united to his body as its form before the question of why his soul is causally paired with his body arises. The very existence of a person's body depends on that person's soul, which grounds the unity of the individual's body as well as its nature. Given that, it is fitting for a person's soul to be causally paired with the person's body.

To summarize, Kim's second and third premises claim that only spatial relations can pair causes with effects, and thus entities must be spatial to stand in causal pairing relations. I've objected by putting forth another relation that is capable of pairing causes with effects. That relation is an en-forming relation. The neo-Thomistic hylomorphist can account for causal pairing by appealing to the en-forming relation the soul stands in to the matter of the body. Therefore, the hylomorphist can justifiably deny premises two and three of Kim's argument.

## OBJECTIONS

Now that my own hylomorphic response to the pairing problem has been given, it is time to address objections to it. In this section, I will evaluate and respond to two objections. First, I will address the objection that my account simply pushes the pairing problem back one level of inquiry. Second, I will evaluate the charge that my hylomorphic explanation of causal pairing, like the Cartesian explanation, presupposes a causal relation. Surely these two objections are not the only objections to my account. Yet for the sake of brevity, I have chosen to focus on these two objections (which are the most threatening objections that have been brought to my attention) in order to provide sufficient responses without taking up too much space.

### First Objection: Pairing Problem Reduxed

The most common objection to my hylomorphic explanation of causal pairing is the charge that it simply pushes the causal pairing problem back one level of inquiry. According to this objection, my account explains why one's soul is causally paired with her body, but now we're left wondering why one's soul en-forms her body. We started with a causal pairing problem and now we are left with an en-forming pairing problem. At least that's the claim of this objection that charges my account with reduxing, or pushing back, the problem I set out to explain. My response is as follows.

To begin with, it is important to notice that the initial question differs from the follow-up question. At the outset, the question is: What explains why soul $X$ is causally paired with body $Y$? My hylomorphic answer is that soul $X$ is causally paired with body $Y$ because $X$ en-forms $Y$. Given this, a different question arises: What explains why soul $X$ en-forms body $Y$? In light of my response to the causal pairing problem, this follow-up question arises. Notice, however, that the follow-up question is not the same as the initial question. The new question pertains to Aristotelian metaphysics generally, since all material substances consist of en-formed matter. The initial question, however, is germane to philosophy of mind. Since the two questions being asked at each level of inquiry are different questions that address different issues, it's not true that the initial question is simply being pushed back. Rather, a different question comes into view in light of my hylomorphic explanation of causal pairing.

One might retort that it is equally troubling that my account leads to a further question that differs from the initial one. However, the fact that my hylomorphic explanation does so is not necessarily problematic either. For such is true of most good explanations. When explanans explain an explanandum, we often wonder what explains the explanans. For example, when we discovered that oxygen gets to our working muscles because red blood cells carry oxygen to them, we didn't say "okay that's the end of our inquiry, let's stop asking questions." Rather, our inquiry moved forward to investigate further questions, such as: Why do red blood cells, rather than other cells, do such? Given this, we could say that the physiologist who discovered the initial fact about red blood cells progressed our inquiry. It is not a problem that the physiologist's explanation raised further questions requiring further research. (Such is considered by many to be a theoretical virtue called *fertility*.) Likewise, the fact that my explanation brings to the fore a new question is not necessarily a mark against it. Many good explanations lead us to further questions.

Secondly, the question my hylomorphic explanation of causal pairing raises is a metaphysical question that all Aristotelians must deal with. The question "what explains why soul $X$ en-forms body $Y$" is difficult to answer. However, anyone who thinks material substances consist of form and matter will have to wrestle with this question (and other related questions) as it pertains to the form and matter of any hylomorphic material substance (cf. Lowe 2012, p. 235; Marmodoro 2013; Rea 2011). For example, we could ask what explains why form $P$ en-forms tree $Q$, or what explains why form $R$ en-forms frog $S$, and so on. As long as you have a form en-forming matter, you can ask what explains that fact. Therefore, the question my hylomorphic account leads to is a metaphysical question pertaining to standard Aristotelianism. It is every Aristotelian's problem—it's not a problem unique to my hylomorphic account of causal pairing.[19] And lest one is tempted to think that makes it worse, it should be acknowledged that all competing metaphysical views have difficult metaphysical questions to answer. So unless one simply doesn't subscribe to any view that entails metaphysical doctrines, they will also have to face challenging metaphysical questions.

Thirdly, the question "what explains why soul $X$ en-forms body $Y$" can evoke a misleading caricature of the human person as a mereological aggregate consisting of two parts that are somehow united (cf. Marmodoro 2013, pp. 5–6). Such a caricature is *not* consistent with the neo-Thomistic hylomorphic view I am proposing. It is worth reiterating that the body, according to the hylomorphic view I am advocating, is not an entity that is united to the soul that is its form. There are not two substances being united. Rather, there is one material substance and an aspect of that substance is the body, which is en-formed matter. Recall that a human soul is a unique substantial form in the sense that it's a substance in itself and in its natural state it en-forms matter; the en-formed matter is the body. Given that, we might say that the soul, in its natural state, is a substance that is an *en-mattered form*.[20] An "en-mattered form" is simply a form that is en-forming matter. It is important to see that the body is not an entity in itself, which the soul then en-forms. On this view, there is one material substance, and the body exists as an aspect of that substance.

Since the body exists only as an aspect of the human person and it's not an entity in itself without the soul that en-forms it, the question "what explains why soul $X$ en-forms body $Y$?" is misleading. After all, there is not a self-existent substance called the body that the soul then en-forms. Instead, there is en-formed matter, and that is the body. The body is not a body without the form. So whenever you are talking about the body, you are talking about matter en-formed by the soul.

Let's consider an imperfect example, which involves a different type of form, but is nevertheless illuminating. The statue *David* could not be what it is without the form that is the shape of the statue.[21] The statue does not exist apart from its shape, so to ask why it has its shape is like asking why it is what is. Given this and that every explanatory trail must stop somewhere, lest an infinite regress ensue, the Aristotelian can justifiable say that it's just a brute fact that the statue *David* has the shape it has and that there's no natural physical or metaphysical explanation. Likewise, since a human body would not be what it is devoid of the soul that en-forms it, the same could be said of it and its form. For without the soul, there is no body to speak of according to hylomorphism. Therefore, it seems legitimate for the hylomorphist to claim that natural explanations stop here, and it is a brute fact that a particular soul en-forms a particular body.

Finally, with the above having been said, let me note that the hylomorphist can give a supernatural explanation of why a particular soul en-forms a particular body if she endorses theism and one element of Aquinas's theology of divine simplicity.[22] Aquinas's explanation of the doctrine of simplicity implies that there are various facets of this doctrine (see Aquinas, ST 1a 3.1-8). Nevertheless, it is clear that Aquinas understood simplicity to at least include the idea that God is not composed of form and matter. It is this idea that I wish to capitalize on here.

In his "Treatise On the One God," Aquinas writes, "it is impossible that God should be composed of matter and form" (Aquinas, ST 1a 3.2c). He thinks God is "of His essence a form; and not composed of matter and form" (Aquinas, ST 1a 3.2c). One might think that if God consisted of form and matter something would have to explain why God's form en-forms the matter it does (cf. Aquinas, ST 1a 3.7c). But given that God is not composed of form and matter, no such explanation is needed. Thus, the explanatory trail of all hylomorphic objects can ultimately stop with God. Since God is not a hylomorphic object requiring the same type of explanation, it is possible that God's creative and sustaining will provide the ultimate explanation of why the form of any hylomorphic object en-forms the matter it does. The idea is that God, who is simple, willed and continues to will that each particular form shall en-form the matter it does with regards to all hylomorphic objects.

One might object: "If you're going to appeal to God, why not appeal to God directly to explain causal pairing relations?" My answer is that such an appeal would be premature and therefore unjustified. According to theism, God is the ultimate explanation of the universe and all things in the universe. But that universe, which God created and sustains, is filled with natural explanations of natural phenomena.[23] Given that, one should appeal to natural explanations insofar as there is reason to think there could be such explanations. However, it is reasonable to appeal to a supernatural[24] explanation when it is reasonable to think there can't possibly be a natural explanation. And since form and matter are most fundamental in the Aristotelian's ontology, there is nothing more fundamental to appeal to as a natural explanation. Therefore, the theist can justifiably appeal to the supernatural explanation of God's creative and sustaining will to account for why forms en-form the matter they do. Granted, materialists won't find this line of thought convincing. Nevertheless, if there are hylomorphic objects, it seems to me that a metaphysically simple God provides a sufficient explanation of why forms en-form the matter they do.

To summarize this subsection, I have addressed the most common objection to my hylomorphic response to the causal pairing problem, which claims that my response simply reduxes the problem. First, I pointed out that the follow-up question my account brings us to is a different question than the initial question, and good explanations often lead to new questions. Then I mentioned that the follow-up question my account raises is germane to standard Aristotelian metaphysics, and therefore it's not a problem unique to my mind-body hylomorphic position. Before making my final point, I clarified that on hylomorphism the body doesn't even exist devoid of a form, and therefore it's fair to say there is no further natural explanation of why a form en-forms the body it does. Lastly, I offered a supernatural explanation that exploits the theological doctrine of divine simplicity. Let us now consider a second objection, which might be the strongest objection to my hylomorphic account of causal pairing.

## Second Objection: Presupposed Causal Relation

Recall that a key charge against the Cartesian dualist is that she can't explain mind-body causal pairing by appealing to mind-body unity since she explains such unity via a causal relation. Such an explanation would presuppose and depend on the very relation it is supposed to explain (see Kim 2005, p. 77). The same problem besets my hylomorphic explanation of causal pairing, according to this second objection. After all, says the objector, the en-forming relation that I appeal to in order to explain causal pairing is itself a causal relation.

The objection is straightforward and can be described as follows. My hylomorphic account says soul $X$ is causally paired with body $Y$ since $X$ en-forms $Y$; therefore, it hinges on an en-forming relation that is itself a causal relation. Since the en-forming relation appealed to in order to explain the causal pairing relation between $X$ and $Y$ is itself a causal relation, it must presuppose a causal pairing relation which it is supposed to explain. Therefore, according to this line of reasoning, my account falls prey to the same charge leveled against Cartesian dualism because it too presupposes the very pairing relation it allegedly explains. That's the objection, and it faces a fatal dilemma.

Either the key premise—the en-forming relation is a causal relation—is false, or else the objection commits the fallacy of equivocation. Regarding the former possibility, if efficient causation is the only type of causation, the key premise is false. On the other hand, if the premise is true, then there are at least two types of causation and two corresponding senses of "cause" that the objection equivocates between. Either way, the objection is fatally flawed. Let's consider each horn of the dilemma in turn, beginning with the possibility that the key premise is false.

In Aristotle's day, he could claim that "things are called causes in many ways" (Aristotle, *Phys.* 195a 4). Hence, his infamous four causes: material cause, formal cause, efficient cause, and final cause.[25] Most moderns, however, disagree with Aristotle's conception of causation. Although an Aristotelian view of causation was commonly held prior to the seventeenth century, during the time of René Descartes and Francis Bacon a shift pertaining to our understanding of causality began to take place (see Gilson 2009, pp. x–xi). Today there is a general conceptual commitment in the minds of most moderns that only efficient causes are causes (Marmodoro 2014b, p. 221). The modern view of causation denies formal causality.

The modern view of causation clearly entails that the key premise of this second objection—the en-forming relation is a causal relation—is false. After all, if the en-forming relation is a causal relation, it's not one of efficient causation, but rather one of formal causation. I will soon clarify the distinction between efficient and formal causality. Yet, for now, the main point is: given the modern view of causation, the en-forming relation cannot be a causal relation since causal relations must be efficient causal relations according to the modern view, not formal causal relations. And if the en-forming relation is a causal relation, it is one of formal causation, not efficient causation. Simply put, if the modern view of causation is correct, the en-forming relation is not a causal relation and the key premise of this second objection is false.

To clarify the difference between efficient causality and formal causality, it will be helpful to consider an example of a particular human embryo $E$ coming into existence. To make things easier we can assume, contra Aquinas, that a human soul is present from the moment of conception.[26] So in this example we're assuming that $E$ is en-formed by a human soul from the moment it begins to exist. If we wished to give an Aristotelian explanation of $E$'s existence, we would want to know *what* it is, what it's *composed of*, what *brought it about*, and what it's *for* (Shields 2014, section 7). These aspects of our inquiry would approximate to Aristotle's four causes. Since we need only clarifying the distinction between two types of causes—formal and efficient—I will focus only on "what it is" and "what brought it about."

The formal cause of $E$ explains what $E$ is, whereas the efficient cause explains what brought $E$ about. The efficient cause is a sperm fertilizing an egg. Had that not happened, $E$ would not have come into existence. But without the formal cause $E$ would not exist as the type of thing it is. The formal cause is what explains why $E$ is a human embryo. And what explains why $E$ is a human embryo is that it's en-formed by a form of a particular type, namely the form of a human person—that is, a human soul.

The formal cause explains why $E$ is a human embryo since the form of the embryo grounds its unity and essence, and thus its existence. The form grounds the essence of the embryo since the essence is determined by the type of form that en-forms it. Accordingly, the characteristic physical properties the embryo develops depends on the form, but the development happens via the means of biological causes that do the efficient causal work, which the form does not do (see chapter 7, section "Biological Regularities"). The form grounds the unity of the embryo, as all the material parts of the embryo depend for existence on the same form. Such unity of a substance that's grounded by its substantial form is very different than, say, the unity of a group of pencils united by a rubber band squeezing them together. Such squeezing would be efficient causation, not formal causation.

Many moderns will hear my explication of formal causation and say, "that's not causation!" I agree, if "causation" merely means "efficient causation." Hence I described the en-forming relation as a non-causal explanatory relation of grounding in the previous chapter (see chapter 5, section "En-forming Relation"). But given that the en-forming relation is not a causal relation, the key premise of the second objection is false and consequently the objection fails. That's the first horn of the dilemma that the objection faces.

Perhaps, however, some will think there are multiple types of causation and formal causality is one type of causality in addition to efficient causality (e.g. Ellis 2009, pp. 77–78). If that is true, the objection equivocates between two senses of "cause," the two senses being *efficient cause* and *formal cause*. This is the other horn of the dilemma.

If it's true that the en-forming relation is a causal relation, it is a relation of formal causality, not efficient causality (cf. Marmodoro and Page 2016, p. 16). Therefore, if my account appeals to a causal relation to explain a causal relation, then it appeals to a relation of formal causality to explain a relation of efficient causality. As a result, it's false to claim that my account presupposes what it is meant to explain, for it presupposes a different type of causal relation than the one it explains. The objection only appears to be successful because it equivocates between two senses of "causal relation."

To summarize, this second objection faces a fatal dilemma. If it's true that there is only efficient causation, which is what the modern view of causation says, then the key premise of the objection is false. On the other hand, if the key premise is true, then the objection fallaciously equivocates. Either way, the objection is fatally flawed.

## CONCLUSION

The causal pairing problem is dualism's chief problem pertaining to mental causation. In this chapter, I have tried to show that there is a substance dualist position—neo-Thomistic hylomorphism—that provides a solution to the pairing problem. According to the hylomorphic solution I've proposed, a person's soul is causally paired with her body because her soul en-forms her body. So, for example, Hillary Clinton's soul stands in a causal pairing relation with her body because her soul stands in the more fundamental en-forming relation to her body. Given that Clinton's soul is the form of her body, it's fitting that her soul is causally paired with her body, rather than, say, Barack Obama's body. In the next chapter, neo-Thomistic hylomorphism will be applied to neural correlates of consciousness.[27]

## NOTES

1. Henceforth, I will often refer to neo-Thomistic hylomorphism simply as "hylomorphism" for the sake of brevity.
2. See chapter 4 for a detailed explication of the pairing problem.
3. See chapter 4 section "Reconstruction of Kim's Argument" for the full argument with sub-premises.
4. Italics mine.
5. Some hylomorphists think the body is a mode of the soul rather than an aspect of the soul (e.g., Moreland 2015, p. 201). This response could be rearticulated accordingly.
6. See chapter 5.
7. This mirrors Kim's (2000, Ch. 2) argument against nonreductive physicalism.
8. Italics mine. The hylomorphist who responds to Kim's argument by way of the first response considered above might want to emphasize the words "involving distinct objects." Since the hylomorphist could agree with Kim on this score and go

on to point out that according to the hylomorphic view a person doesn't consist of two distinct objects, but rather two distinct aspects of one substance.

9. Emphasis mine.

10. As suggested via an endnote in chapter 3, section "Reasons to Think Closure is False," the following are instructive sources regarding the rationality of theism: Walls and Dougherty (2018), Craig and Moreland (2009), Plantinga (2000), Moreland et al. (2013), Moreland (2008), Swinburne (1996), Copan and Craig (2018a; 2018b), and Copan and Taliaferro (2019).

11. Some might quibble with this position on the basis that God is omnipresent and go on to infer that God must therefore be spatial. But even if such were so, it wouldn't threaten the nonspatiality of God prior to the creation of all things including space.

12. Cf. Plantinga (1984, 2000, 2007).

13. Kim (2005, p. 170) does say "ordinary sensory concepts, like 'pain' . . . have motivational/behavioral aspects in addition to qualitative/sensory aspects" and that the former aspects are functionalizable and thus reducible. But this qualifier doesn't negate his prior claim: "Pain as a sensory quale is not a functional property" (Kim 2005, p. 169). Thus, he thinks that a state of pain, as a quale, is a part of the mental residue that's not functionalizable and, therefore, irreducible and nonphysical.

14. If causation depended on our understanding of it, then there could not have been causation before humans came into existence and became capable of understanding causation. It would follow that human origins could not have a causal explanation, given that there could be no causation predating human understanding of causation.

15. Moreland (2018, p. 113) commends the type of response I'm advocating here. Although he doesn't develop the response, I find his commendation confirming.

16. For a description of the en-forming relation, see chapter 5 section "En-forming Relation."

17. It should be noted that the Cartesian might think that the causal pairing relation is fundamental and therefore doesn't need a natural or metaphysical explanation. In that case, she wouldn't give one and thus wouldn't fall prey to Kim's following refutation.

18. See Stump (2003, p. 200).

19. I'm indebted to Nikk Effingham for bringing this point to my attention.

20. I'm borrowing the term "en-mattered form" from Marmodoro (2013, p. 5).

21. As previously mentioned (see chapter 5 section "Forms"), the common example of a statue and its shape has shortcomings when it comes to explaining the nature of a substantial form and a material substance. Nevertheless, it can help us understand the specific point made here.

22. For an overview of Aquinas's understanding of divine simplicity, see Stump (2012).

23. "Natural explanations" here refers to physical and metaphysical explanations.

24. By "supernatural explanation" I do not mean an explanation that violates the natural order. Rather, I am referring to an explanation that transcends the natural order

and explains it. A supernatural explanation of the initial singularity, for example, need not in any way violate the natural order of the physical universe.

25. For an introductory explanation of Aristotle's four causes, see Shields (2014).

26. Aquinas thought an embryo first has only a sensory soul and then later comes to have a human intellective soul (see Aquinas, ST 1a 76.3 ad 3).

27. An antecedent version of this chapter was published as a research article in a special issue of *Synthese* entitled "Form, Structure and Hylomorphism," edited by Anna Marmodoro and Michele Paolini Poaletti. Reprinted by permission from Springer Nature: Springer Nature, Synthese (Owen, Matthew, "Circumnavigating the causal pairing problem with hylomorphism and the integrated information theory of consciousness"). Copyright © 2019. https://doi.org/10.1007/s11229-019-02403-6

*Chapter 7*

# The Mind-Body Powers Model of NCC

Since discussing neural correlates of consciousness in chapter 2, we have covered much terrain. A discussion of problems pertaining to mental causation occupied our attention in chapters 3 and 4. After introducing neo-Thomistic hylomorphism in chapter 5, I argued in chapter 6 that it provides a solution to the causal pairing problem. Now in this chapter, our focus will return to neural correlates of consciousness (for brevity NCC), upon which rests the second chief objection to dualism.

Recall that in chapter 2 I introduced the subject of NCC, discussed what exactly they are, how we identify them, and, most importantly, what they imply. As discussed in some detail, neural correlates themselves don't entail any particular view of the mind (see chapter 2, section "Implications of NCC"). Given that, one cannot undermine dualism by simply pointing to neural correlates. Philosophical argumentation is also needed in order to show that the correlations imply dualism is false; hence the simplicity argument for the identity theory. In the remainder of the second chapter, I presented and critically analyzed the simplicity argument and found it insufficient to undermine dualism.

My aim in this chapter is to show that neo-Thomistic hylomorphism is not only consistent with NCC but also provides a good explanation of NCC. To accomplish this, I will build upon the work of other Aristotelian and Thomistic hylomorphists to construct a model of neural correlates of consciousness. Since I am not the first Aristotelian sympathizer to explore the new world of modern neuroscience, my goal is simply to advance the exploration.[1]

To begin with, in the following section I will give a very brief historical sketch of contemporary discussions pertinent to hylomorphism and neural correlates of consciousness. Although there are both materialist and dualist leaning hylomorphists, my focus will be toward the dualist side in order to clarify the context surrounding my own hylomorphic account of NCC. This first section will also serve to clarify my objective and the nature of my proposed hylomorphic explanation of NCC. The following four sections focus on explicating the conceptual foundation for the proposed model.

While my primary objectives are not exegetical, sections "Aristotelian Powers Ontology" and "Mind-Body Dependence" introduce and explain principles derived from the works of Aristotle and Aquinas. The section "Aristotelian Powers Ontology" introduces relevant ideas regarding Aristotle's ontology of powers. The section "Mind-Body Dependence" discusses the dependence of the mind on the body, a principle evident in Aquinas's "Treatise on Human Nature." Applying principles gained from Aristotle and Aquinas, in the section "Mind-Body Powers" I will explicate what I call "mind-body powers." The explanation of the conceptual framework situating the proposed hylomorphic account is finalized in the section "Biological Regularities," which discusses the grounding for biological regularities. The section "Mind-Body Powers Model" applies the principles clarified in prior sections to construct the proposed account of NCC.

## RECENT HYLOMORPHIC HISTORY

Not long after the contemporary search for neural correlates of consciousness began, David Oderberg (2005, p. 90) claimed: "For the hylemorphic dualist, such correlations are only to be expected since persons as embodied beings require corporeal activity in order to interact with the world."[2] If the mind depends on the body to fulfill the functions of a human person, then there will be neural correlates of mental activity, Oderberg reasoned. Recently, J. P. Moreland has (2016, pp. 116–119; 2018) contributed to this line of thought and argued that a Thomistic-like version of hylomorphism *entails* NCC.

Moreland (2016, p. 116) arrives at his conclusion by appealing to concepts he believes reflect two versions of "Aristotelian-style dualism." One is what Moreland (2016, p. 119) calls *organicism*, according to which the structure and function of body parts are determined by the functional demands of the soul that's "the *first efficient cause* of the body's development."[3] As it stands, this principle is not part of my explanation of NCC in what follows. As I will discuss in the section "Biological Regularities," I agree with Moreland that the soul determines the structure and function of the body's parts; that much is standard Aristotelianism and part of my hylomorphic explanation of NCC. However, I deny the idea that the soul is an efficient cause of the body's development. In the section "Biological Regularities," the reason for my denial will be discussed.

The other principle Moreland appeals to—mind-body dependence—is vital to my own explanation, and I intend to elaborate on it significantly. Moreland (2016, p. 117) appeals to this principle held by the late medieval Aristotelians as it is articulated by Dennis des Chene (2000, p. 71):

> by its nature [the human soul] presupposes union with a body, and moreover with a particular kind of body, a body with organs, in order to exercise all its powers—even reason insofar as reason needs the senses to give it material for abstraction.

The basic idea is that there is a relation of dependence between the mind and the body, and the mind relies on the body to manifest its powers. According to the late Aristotelians, "even the intellect requires, so long as the soul is joined with a body, a certain disposition of the brain" (des Chene 2000, p. 96; see also Moreland 2016, p. 117).

As I will discuss in the section "Mind-Body Dependence," the idea that the body is integral to the soul's functions is evident in Aquinas's writings. The late Aristotelians concurred, and the same is true of many Aristotelian leaning philosophers today. Both John Haldane (1998, pp. 271–272) and E. J. Lowe (2006, pp. 11–19) seem to embrace this idea when discussing the relationship between mental intentionality and neurophysiology.[4] In the context of the contemporary debate about causal overdetermination, the naturalist leaning hylomorphist William Jaworski (2016, p. 281) appeals to this concept to show that mental and bodily causes are both necessary and thus there's no overdetermination (see the section "Mind-Body Powers Model").[5]

According to Oderberg (2005, p. 90) and Moreland (2016, p. 117), this idea that the soul relies on the body to exercise its powers suggests that we should expect neural correlates of consciousness. Moreland (2016, p. 117) moves from the idea of mind-body dependence to the conclusion that NCC are entailed. Oderberg and Moreland's objectors might allege that their reasoning is too hasty since the dependence of the mind on the body doesn't necessarily explain or entail patterned processes at the neuronal level, such as NCC. As Moreland (2016, p. 126) acknowledges, there is more to be said—additional premises are needed, details must be filled in.

I intend to progress this Aristotelian-Thomistic line of thought and to demonstrate that neo-Thomistic hylomorphism can provide a good explanation of NCC. My objective is to propose an explanatory model of neural correlates of consciousness, which I call the *Mind-Body Powers model of NCC*. It includes the mind-body dependence referenced by Oderberg and Moreland that is evident in Aquinas's thought, but the model goes beyond philosophy of mind and is anchored to fundamental metaphysical principles regarding Aristotelian causation. My claim is not that neo-Thomistic hylomorphism entails NCC, but rather that it can provide a good explanation (or model/ account) of NCC. The explanation, however, is not a scientific explanation. It is a metaphysical explanation.[6] It provides a coherent metaphysical framework for understanding NCC that can provide warrant for NCC research and help us coherently interpret NCC data.[7]

The next four sections clarify Aristotelian-Thomistic principles that inform the Mind-Body Powers model of NCC. After the metaphysical underpinnings have been clarified, I will present the model itself. In the following chapter, the model will be applied to the issue of empirically discerning and quantifying consciousness.

## ARISTOTELIAN POWERS ONTOLOGY

When introducing my neo-Thomistic hylomorphic position in chapter 5, I mentioned two desiderata, which are somewhat competing. The first desideratum is that the view be as minimal as possible, in that it commits adherents only to tenets that are essential. I want it to be possible to adopt the view without becoming a full-blown Aristotelian on every metaphysical topic. With that said, the second desideratum is for the proposed hylomorphic view to be as defensible as possible and as capable as possible of overcoming serious objections. Sometimes a more minimal version of a position is more defensible because there is less to defend. At other times, a more robust version is more defensible. One reason for this is that additional tenets can sometimes allow a view to explain more phenomena, giving it more explanatory power.

To overcome the pairing problem in the previous chapter, we only needed the most fundamental tenet of hylomorphism that says the soul is the form of the body. In this chapter, however, I am going to put Aristotelian powers to work in my account of NCC. To do so, I will need to commit to certain Aristotelian concepts regarding powers. In this section, I will clarify and explain such concepts. Throughout this section I will often rely on Anna Marmodoro's (2014a) reading of Aristotle with regards to powers. In due course, the concepts I glean from Marmodoro's interpretation of Aristotle will be applied to my neo-Thomistic hylomorphic account of NCC.

### Powers

Aristotle thought objects have powers. According to Aristotelian metaphysics, a power is the capacity to produce change or undergo change (see Marmodoro 2014a, p. 13).[8] In *Metaphysics* (V 12), Aristotle provides this general definition:

Capacity [i.e. power] then is the source, in general, of change or movement in another thing or in the same thing *qua* other, and also the source of a thing's being moved by another thing or by itself *qua* other.

Aristotle here characterizes a power as a source of change. Notice that this source of change includes that which brings about change and also that which undergoes the change (Marmodoro 2014a, p. 12).

The capacity to bring about change is the *active power*, whereas the capacity to undergo change is the *passive power*. Active powers are what they are due to the type of change they can produce, and passive powers are what they are due to the type of change they can undergo. Both types of powers have two potential states. A power can be manifested, or in other words "activated" or "exercised." Yet a power can exist even when it's not being manifested in a state of potentiality, according to the Aristotelian powers ontology I am endorsing. For example, an ordinary bowling ball has the power to roll whether or not it is currently rolling. When it is rolling, its power to roll is manifested. When it is not rolling, it still has the power to roll, but the power is in a state of potentiality. Because the bowling ball has the power to roll even when it is sitting still on a rack, it makes sense for the bowler to take it off the rack and then try to roll it down the bowling isle.

A unique feature of Aristotelian powers is that active and passive powers depend on one another for their manifestation—active powers depend on passive powers and passive powers depend on active powers (Marmodoro 2014a, p. 13). As Marmodoro (2014a, p. 13) writes, "Aristotle defines an active power as one that exercises its powerfulness on a corresponding passive one." To use Aristotle's well-known example of a teacher and a learner, the manifestation of the teacher's active power to teach depends on the learner's passive power to learn. If the learner does not manifest the passive power to learn, the teacher's active power to teach won't be manifested.

Furthermore, the teacher's active power to teach is manifested, or actualized, in the learner who is taught by the teacher, as Aristotle explains in *Physics*:

> It is not absurd that the actualization of one thing should be in another. Teaching is the activity of a person who can teach, yet the operation is performed in something—it is not cut adrift from a subject, but is of one thing in another. There is nothing to prevent two things having one and the same actualization (not the same in being, but related as the potential is to the actual). (Aristotle, *Phys*. III 3)

The manifestation of the active power to teach is realized in the manifestation of the passive power to learn, and in this way the manifestation of the active power ontologically depends on the passive power (see Marmodoro 2014b, pp. 243–244). Another example that illustrates this point is a hot metal pot's capacity to heat the water it contains, which is realized through the manifestation of the water's passive power to be heated.

Running with this example, I want to make a point about the relevant complexity regarding active and passive powers. To do so, I will make the example more complex but also more realistic. Suppose the metal pot is heated by a hot-stove plate so that the surface of the pot reaches 150°C. Let's also assume that the water in the pot initially had a temperature of 5°C, but since the surface temperature of the pot has risen to 150°C, the temperature of the water rises. In this scenario, which is still relatively simplistic compared to many causal instances in nature, things have become more complex. The hot-stove plate is manifesting its active power to heat the pot. The pot is manifesting its passive power to be heated as well as its active power to heat the water. Additionally, anyone who has ever heated a metal pot with cold water in it knows that it takes longer than heating an empty metal pot because the water acts on the pot as a cooling agent. So the water is also manifesting its active power to cool the pot, which is manifesting its passive power to be cooled. And as the water affects the pot's surface temperature, we can reasonably assume that this cools the hot-stove plate, however minor the cooling affect may be. The picture that has emerged here is one where the stove plate, the pot, and the water are all acting on one another and being acted on by one another. In other words, each item involved is manifesting active powers and passive powers simultaneously, and there is mutual causation taking place.

Is this problematic for Aristotelian powers? Well, as Marmodoro (2014a, p. 41) points out, "Aristotle acknowledges, for example in *Physics* (202a5-12), that in most causal interactions in nature the change is mutual." While our complex example is not simple, it better represents the causal complexity in nature, which is even far more complex than the example illustrates. Yet what is still taking place, despite the complexity, is that things with active powers are producing change and things with passive powers are being changed. And despite the complexity of real causal cases, simplistic examples have their place. Simple examples help us understand the concepts involved, even if they don't accurately capture all the complexity of exactly how they are involved in the reality of nature, given its intricacy. I have already given such simplistic yet effective examples and will continue to do so.[9] The reader can safely assume that the reality of causation is certainly more complex than the examples, while the general concepts being conveyed are thought to nevertheless apply in all the realistic complexity.

To recap, according to Aristotelian metaphysics, entities have powers that are real objective features of the world.[10] Some powers are active and some are passive. An active power is the capacity to produce change, whereas a passive power is the capacity to undergo change. Manifestations of active powers depend on manifestations of passive powers, and vice versa; and particular active powers regularly depend on particular passive powers. Aristotelian powers, as Marmodoro (2014a, p. 32) puts it, "depend for their activation on the activation of their mutual partner-powers." The following tenet offers a succinct summary of the most important points discussed in this section.

> *Interdependent Partner-Powers* (IPP): Things have active and passive powers. The manifestation of an active power(s) ontologically depends on the manifestation of its corresponding passive power(s), and vice versa.

In due course, I will apply this concept regarding corresponding (or, correlated) active and passive powers to neural correlates of consciousness. Yet before doing so, I need to introduce the idea of mind-body dependence and then what I call "mind-body powers."

## MIND-BODY DEPENDENCE

To account for NCC, I will employ another principle that is distinct from IPP but in a sense an application of it to the manifestations of mental and bodily powers. Simply put, the principle is that the soul manifests its powers via macro body parts (e.g., organs) and micro body parts (e.g., cells) that are organized and structured in particular ways to manifest particular powers. This principle has exegetical origins in the works of Aquinas, and this section briefly goes over some key texts to make that evident. However, since my focus is not exegetical, my hermeneutical work will be minimal.

As one reads Aquinas, it is easy to see that he thought powers of the human soul rely on the body. Aquinas (ST 1a 76.8c) makes it clear that powers of the soul are manifested through bodily organs and even located in such organs. Given this, he thought certain powers cease being exercised when the body ceases. According to Aquinas (ST 1a 77.8c), these powers "virtually remain in the soul" but their manifestation depends on the body, "for the action of such capacities occurs only through a corporeal organ." The manifestation of the soul's powers happens through the function of organs, and it is not as though any organ will suffice for the manifestation of any particular power. Rather, specific parts of the body exist to carry out specific powers of the soul and are thus structured accordingly. For example, the power to see depends on the eyes, which are structured for seeing, and the power to choose to run depends on the legs, which are structured for running.

The idea that certain powers of the soul are in and manifested through corporeal organs that are structured to carry out such powers can be applied to the microscopic level of cells as well. Granted, Aquinas did not apply it to the cellular level. But given that he allowed empirical findings to inform his philosophical positions, he could have easily developed this principle to apply to the cellular level if he knew what we know from modern biology. When the principle is so modified in the light of modern biology, it suggests that there will be cells, such as nerve cells (i.e., neurons), that manifest certain powers ultimately belonging to the soul.

So far, so good, when it comes to most powers, those familiar with Aquinas's work might say. Yet, they will quickly point out the fact that rationality is more complicated on Aquinas's view. Like Aristotle, Aquinas considered rationality to be definitive of humanity.[11] He also made a case for the immortality of the human soul that relies on the power of rationality (see Aquinas, ST 1a 75.2c, 75.3c, and 75.6c). He argued that the human soul is immortal and continues to exist after bodily death and before the resurrection of the body because it has its own operation independent of the body—that is, rational cognition. In other words, Aquinas thought the soul is immortal because the manifestation of rationality does not necessarily require a bodily organ (see Pasnau 2002b, p. xvii).

It is easy to misunderstand Aquinas at this point. In light of Aquinas's prevalent acknowledgment of the body's integral role in the manifestation of the soul's powers, one expects to see the same regarding rationality. In several places, however, Aquinas seems to suggest that the body plays no role in rational thought (see, e.g., Aquinas, ST 1a 76.1c and 76.1 ad 1). However, I do not think this is Aquinas's position because in other places he clearly affirms a role for the body in rational thought. For example, Aquinas (ST 1a 84.4c) writes:

> But a body seems necessary for the intellective soul above all for its proper operation, which is to understand. For the soul does not depend on the body for its existence. But if the soul were naturally suited to receive intelligible species solely through an influx from certain separate principles, and if it did not take in species through the senses, then it would not need a body in order to understand and so it would be pointless for it to be united to its body.

This passage makes it hard to deny that Aquinas thinks the body plays a role in the intellectual act of understanding.

In another place Aquinas even seems to pinpoint the brain as the organ of thought. After writing, in question 78, article 3, that "capacities do not exist on account of organs, but organs on account of capacities," in the next article Aquinas (ST 1a 78.4c) claims that physicians have assigned a definite organ to "particular reason," that organ being "the middle part of the head." Robert Pasnau (2002a, p. 283) comments that Aquinas followed the contemporary science at the time in holding that the four internal senses (i.e., common sense, imagination, cogitative power, and memory)[12] are in the brain. Given that, it is unsurprising that Aquinas thought that damage to this organ so relevant to understanding would impede understanding (Aquinas, ST 1a 84.7c).

How can we make sense of Aquinas apparently contradictory affirmations? On the one hand, it seems that he is saying human rationality is exercised independently of the body. Yet, on the other hand, he seems to affirm that the body plays a role in rational thought. I think Aquinas's view can be summarized as follows: The soul's capacity of rational thought is capable of operating independently in the sense that it does not metaphysically necessitate a bodily organ, although its natural operation uses a bodily organ, which is required for the natural operation.

This reading finds textual support where Aquinas says that the soul thinks through what he calls "phantasms" (or sense images) that depend on the body. Toward the end of his treatise on human nature in the *Summa*, Aquinas (ST 1a 89.1c; cf. 75.7 ad 3) writes:

> Therefore with respect to the mode of existence by which the soul is united to a body, the appropriate mode of understanding for the soul is to turn toward the phantasms of bodies, phantasms which exist within bodily organs. But once it has been separated from its body, the appropriate mode of understanding for the soul is to turn toward intelligible things straightaway—just as is appropriate for other separate substances. So turning toward phantasms is, for the soul, its natural mode of understanding, just as being united to a body is natural. But being separated from its body is foreign to the character of its nature, and understanding without turning toward phantasms is likewise foreign to its nature. So it is united to a body in order to exist and operate in keeping with its nature.

There is a sense in which the defining capacity of humanity—rational thought—does not necessitate the body for its operation, and there is a sense in which it does. It is metaphysically possible for the soul to manifest rational thought apart from the body. However, the soul naturally en-forms a body through which the human soul's defining capacity is naturally manifested, in a mode involving phantasms (or sense images). In sum, the human soul's capacity of rational cognition is *naturally* manifested in a way that relies on sense images which depend on the body, yet it's *metaphysically* possible for it to be manifested in an alternative way apart from the body.

Clearly, there is much more to say about how to interpret Aquinas on this issue. Since that is not my focus, however, let me simply clarify the principle I have gleaned from Aquinas's thought and apply in my account of NCC. That principle is that particular mental powers of the soul rely on certain capacities of the body, which are manifested by body parts sufficiently structured to manifest such powers. This idea coupled with IPP informs the following tenet that I employ in my account of NCC:

> *Mind-Body Powers* (MBP): There are mental powers of the soul and physical powers of the body that are interdependent partner-powers. The natural manifestation of an active partner-power(s) (whether mental or physical) ontologically depends on the manifestation of its passive partner-power(s) (whether mental or physical), and vice versa.

Before going on to apply this tenet regarding mind-body powers to neural correlates, it will be helpful to further explain the nature of such powers in the following section.

## MIND-BODY POWERS

According to neo-Thomistic hylomorphism, human persons have mental powers and physical powers of the body that are interdependent partner-powers. I've dubbed such powers "mind-body powers." My model of NCC critically relies on these powers, and therefore it is important to accurately understand them in order to understand the model. This section further elucidates the nature of these mental and bodily powers that are interdependent partner-powers.

### Interdependence

It is easy to imagine powers of the body that can be manifested apart from mental powers. For example, my stomach is digesting my lunch without the exercise of any mental powers. We can also imagine a mental power that manifests without depending on a bodily power. For example, we can conceive of a disembodied mind perceiving God in what Christian theologians refer to as the intermediate state after bodily death and before bodily resurrection. Such powers are not my present concern. Rather, I am interested in powers of the body and mind that naturally require mutual manifestation, or co-manifestation. In other words, my focus is specifically on the mental and physical partner-powers that I call mind-body powers.

To illustrate the concept of mind-body powers, let's use an example of a voluntary act of running undertaken by Olympic gold medalist, Allyson Felix. We can keep things simpler by stipulating that the voluntary act of running relies on the mutual manifestation of the following partner-powers. Felix manifests a mental active power: (*M-Power*) choosing to run. As a result, she manifests passive bodily powers: (*B-Power*) neurons fire, muscles contract, and her legs stride one in front of the other. In actuality things are much more complex, especially on the physiological level. (The bodily physiological powers will include countless manifestations of active powers and passive powers as well as powers that are both active and passive.) But our example need not include all the nuances to illustrate the key idea regarding mind-body powers. That is, both *M-Power* and *B-Power* are needed for this voluntary act of running undertaken by Felix.

Because Felix chose to run, manifesting *M-Power*, neurons start firing and Felix's legs begin striding forward, manifesting *B-Power*. Since she chose to run the manifestation of *B-Power* is part of a voluntary act of running. In this way the manifestation of *B-Power* ontologically depends on the manifestation of *M-Power*, choosing to run. Yet it is also true, according to my account, that *M-Power* could not be manifested without *B-Power*. Granted, Felix could manifest the power to intend to run without the correlated *B-Power*, but intending to run is not the same as choosing to run. One can only choose to do something that she has the choice to do, and she only has the choice to do $x$ if she can do $x$. *B-Power* makes it possible for Felix to run and thus possible for her to choose to run. So in this way the power to choose to run—that is, *M-Power*—ontologically depends on *B-Power*.

Aquinas made a similar claim. According to Aquinas (ST 1a 76.4 ad 2), the soul moves the body "through its potential for producing movement, the actualization of which presupposes a body." The soul has the power to move the body, but this power cannot be manifested without a body. Notice, however, that Aquinas claims that the power to produce bodily movement presupposes *a* body, not this particular body or that particular body. A body will do. Felix need not have the exact physiological makeup that she in fact has in the actual world in order to have and manifest *M-Power*. She could have had a different physiological makeup with different physical powers that still resulted in legs striding forward as a result of the manifestation of *M-Power*. This is important because it allows for multiple realizability.

Our example involves an active mental power. However, we could have given an example with a passive mental power, such as a pinched nerve being the active power that changes one's felt experience to a state of pain. There can be active mental powers and active bodily powers. Likewise, there can be passive mental powers and passive bodily powers. Regardless of whether an active power is physical or mental, its manifestation ontologically depends on the manifestation of its correlated passive partner-power. Likewise, mental and physical passive powers depend on correlated active powers. These interdependent mental and physical bodily partner-powers are mind-body powers.

## Ontological Extension

It's tempting to picture in our minds an external relation relating an active power to its correlated passive power, but such a depiction is inconsistent with an Aristotelian metaphysics of causation. As Marmodoro (2014b, p. 244) explains:

> The causal interaction of the active and the passive powers is not reified by Aristotle as a relation, but as an ontological extension of the agent onto the patient. Aristotle does not posit a relation between active and passive powers to explain the mechanism of causation, but treats the active power as "extending" onto the passive one, not through a relation but by "spreading itself" onto the patient—by making the patients constitution part of the agent's own constitution; by having the patient as the ground of realization of the agent's own causal power.

The manifestation of the passive power is a constituent of the active power's manifestation. The manifestation of the active power comes to be completely realized in the manifestation of the passive power.

Recall Aristotle's example of the teacher and the learner. For the teacher to manifest her power to teach, the learner must manifest her power to learn. If the learner does not learn, the teacher does not teach. On the other hand, when the learner learns from the teaching, the teacher's act of teaching is partly realized in the learner's act of learning. We might say that the teacher's active power to teach is partly realized in the manifestation of the learner's passive power to learn. The teacher's power to teach is not reducible to the learner learning, but the manifestation of the learner's power to learn is a constituent of the manifestation of the teacher's active power to teach.

Returning to our example of Allyson Felix's voluntary act of running, her act of choosing to run is realized in neurons firing, muscles contracting, and legs striding forward. That is, the manifestation of Felix's *M-Power* is realized in the manifestation of *B-Power*. That does not mean *M-Power* is reducible to *B-Power*, but it does mean that the manifestation of *B-Power* is a constituent of the manifestation of *M-Power*. Since *B-Power*'s manifestation is a constituent of *M-Power*'s manifestation, the manifestation of *M-Power* ontologically depends on the manifestation of *B-Power*. And as I will discuss next, the manifestation of mental powers requiring the co-manifestation of body powers depends on physical properties.

## Requisite Physical Properties

Birds often dart past my office window, making it easy for an adrenaline junky like myself to covet their capacity to fly. Despite the fact that I would like to fly like a bird on my own strength without the aid of technology (such as a plane or a wingsuit), I cannot do so. Such is impossible for me, and the reason for this is obvious. Birds have a different type of body with different body parts capable of manifesting different bodily powers. Moreover, certain actions require bodily powers manifested by body parts with sufficient physical properties. I do not have the physical properties sufficient for manifesting the power to fly.

A critical tenet of the hylomorphic view I'm advocating is that certain types of mental powers require certain types of bodily powers, and vice versa. Moreover, certain types of bodily powers require certain types of bodily features. In other words, certain types of mental powers cannot be naturally manifested apart from the manifestation of certain types of bodily powers that require certain types of physical characteristics in the body. The manifestation of Felix's mental power to choose to run requires the bodily power to stride forward with one's legs. And this bodily power to stride forward requires a type of physical bodily feature, namely legs. Felix could have various types of legs and still run, but she does need legs of some type.

Aquinas thought that the human soul has a variety of powers, and therefore the human body must have a variety of powers so that the soul's various powers could be manifested. In question 76, article 5, of the first part of the *Summa Theologiae*, Aquinas addresses the question: What type of body should have the intellective human soul as its form? In this context, Aquinas (ST 1a 76.5 ad 3) writes the following about the human, intellective soul:

> For although it [i.e. the intellective soul] is one in essence, nevertheless, due to its perfection, it has *multiple powers*. Hence for its various operations it needs *various dispositions in the parts of the body* to which it is united. This is why we see that perfect animals have a greater *diversity of parts* than do imperfect animals.[13]

Notice that Aquinas moves from the powers of the soul to the powers of the body to the parts of the body. To manifest its various powers, the soul needs various body parts with the physical properties sufficient for exercising various bodily dispositions, or powers.

If the powers of the human soul could be manifested with the co-manifestation of just any body power using just any body part, then a body with little or no variety of parts would do. There would be no need for the human body's diversity of parts. Yet, according to Aquinas, the human soul requires a body of a certain type, namely one with a variety of biological parts capable of manifesting various bodily powers. Such a body is needed because the soul's various powers depend on the co-manifestation of particular powers of the body manifested by various body parts with sufficient physical properties to manifest the various powers.

This line of thought was clearly applied by Aquinas on the macro level of bodily organs. For example, Aquinas (ST 1a 54.5c) claims: "In our soul there are certain powers whose operations are exercised by means of corporeal organs; such powers are acts of sundry parts of the body, as sight of the eye, and hearing of the ear."[14] Here Aquinas applies to the level of organs the idea that the soul's powers rely on bodily dispositions manifested by certain types of biological body parts with sufficient physical characteristics (see also Aquinas, ST 1a 76.8c, ad 4, ad 5). The same concept can be applied not only to the macroscopic level of organs but also to the microscopic levels of human biology as well.

Due to technological advances since Aquinas's day (1225–1274 AD), we have learned much about the human body beyond the level of organs. In light of modern biology, we can apply Aquinas's line of thought about requisite physical properties to micro levels of bodily composition. From macro to micro levels, we can infer, body parts with certain physical properties or types of properties are needed in order to manifest certain bodily powers that are interdependent partner-powers of mental powers. The fundamental idea is that mind-body powers require certain types of physical properties at every level of bodily composition, from the macro level of bodily organs to the most micro, subcellular level. This idea, which I will apply to cognitive neurobiology via the Mind-Body Powers model of NCC, can be summarized in the following principle:

*Requisite Physical Properties* (RPP): From macro to micro levels, biological body parts must have sufficient types of physical properties in order to naturally manifest the bodily powers that co-constitute mind-body powers.

The requisite physical properties can pertain to anything physical, from physical structure to physical complexity, to electrical charge, to a certain chemical composition, and so forth. The physical properties might be something very specific or a general type of physical characteristic.[15] Of course, we will want to know what physical functions and properties manifest specific bodily powers when a particular mental power is manifested, and why. This is where the empirical sciences, such as cognitive neuroscience and neurophysiology, can be invaluable.

In any event, the key takeaway from this section is that the natural manifestation of mental powers requires the co-manifestation of certain bodily powers that require certain types of physical properties at every level of the body's composition. This concept, which I think is an implication of Aquinas's thought, has explanatory power when it comes to accounting for neural correlates of consciousness. Yet before applying it to neurobiology via the Mind-Body Powers model, we must specify one last principle undergirding the model. This final principle pertains to the form of the body and biological regularities.

## BIOLOGICAL REGULARITIES

One thing about NCC that invites an explanation is the regularity with which particular neural processes correlate with certain types of mental states throughout a species. Consider again the well-known (though factually inaccurate) example of firing C-fibers and the mental state of pain. Assuming for illustration that firing C-fibers are an NCC of pain, it is not just that C-fibers fire in Mary's brain when she is in a state of pain, but that C-fibers fire in John's brain, Eleonore's brain, Cecilia's brain, Juan's brain, and Aryn's brain when each is in pain, and the list goes on and on. There is a regular consistency of the NCC throughout the human species. Such consistency is a standard feature of NCC and entailed by what it means to be an NCC.[16] It is such consistency across a species that makes it possible to map a particular type of brain, such as the mouse brain or the human brain.[17]

The Mind-Body Powers model of neural correlates explains the consistent regularities of NCC across the human species by appealing to the hylomorphic idea that all humans have the same type of form. That form, according to neo-Thomistic hylomorphism, establishes the biological regularity of physical characteristics in the human body across the human species at the macro and microscopic, as well as the submicroscopic, biological levels. In this section, I will briefly explain this idea and, in doing so, finalize the conceptual foundation for the model of NCC proposed in the next section.

As mentioned above, Moreland (2016, pp. 116, 118–119) argues that his Thomistic-like dualism entails NCC by appealing to organicism. According to Moreland's (2016, p. 119) description of organicism, the structures and functions of body parts are determined by the functional demands of the soul, which is "the *first efficient cause* of the body's development."[18] There is significant common ground between Moreland's view and my own, but also an important difference.

Moreland and I agree that the soul is the form of the body, and as such it ultimately determines the biological structures and functions of the body's various parts. We also agree that the functional demands of the soul play a role in this determination (see Moreland 2016, p. 119). Furthermore, I agree with Moreland's (2018, pp. 105–108) claim that the form provides the information for the body's organization and structure, as a blueprint analogously provides the organizational and structural plan of a building. We also appear to agree that the form of the body determines the information in DNA, which guides the development of the body (see Moreland 2018, pp. 105–108). However, I disagree with Moreland's (2016, p. 119) claim that the soul is the first efficient cause of the body's development.

On my view, the form ultimately grounds the biological regularities across a species, including standard biological structures and functions, but this does not involve the form as an efficient cause of the body's development. I would echo the words of John Haldane (1998, pp. 275–276):

> I wish to maintain that form may be a determinant of the substantial nature, including the characteristic activities, of a substance without that being a matter of efficiently constraining the location and behavior of basic particles.... What form brings is order, but it does not do so by pushing things this way or that. Its existence is testified to not by force detectors but by the fact that what exists, and how existents act, exhibit natural order.[19]

Likewise, I am inclined to think the substantial form determines the standard biological order across a species and yet this does not involve the form as an efficient cause.

There are countless biological regularities in our world. Dogs regularly have four legs. Tadpoles regularly become frogs. North American bull elk regularly have antlers, which they regularly shed annually. The list of biological regularities is endless. The type of biological regularity we are concerned with is the physical properties had by biological human body parts that are the same across the human species from macro to micro levels of bodily composition. An example of a macro regularity would be the sameness in structure and function of the human heart throughout the human species. An example of a micro regularity would be the common information encoded in DNA throughout the human species.

According to neo-Thomistic hylomorphism, macro and micro biological regularities across a species are ultimately grounded in the fact that the members of the species have the same type of form. Substantial forms ground the essential properties and powers of the organisms they en-form, determining the developmental parameters of the organism (cf. Pruss 2013a, p. 131). This guides the patterns of biological development that produce organized structures with standard functions as well as other physical characteristics via biological causes, which are the efficient causes of such development. On one explanatory level, there is the form that grounds the standard biological properties an organism comes to exemplify. On another level, there is the physical processes of biological development that are the efficient causes that explain how the biological properties come to be. As Aquinas (ST 1a 76.5c) clarified, "the reason why matter is such as it is must be drawn from the form, not vice versa." The form explains the reason why the physical properties of the body are what they are, but it is biological processes that explain how (in the sense of efficient causation) they come to be.

One might wonder how all this is consistent with neuroplasticity, which is the nervous system's capacity to reorganize its neuronal structures, functions, and connections in response to internal and external stimuli (see Cramer et al. 2011). Due to neuroplasticity, the human brain can undergo traumatic injuries affecting its standard structure and function and yet adapt so as to retain or regain mental capacities that typically rely on the damaged brain areas that are altered as a result of injury (see Cramer et al. 2011; Goodrich et al. 2013; Munoz-Cespedes et al. 2005). Such a possibility might seem incompatible with the idea that the form of the human body determines that its parts will have particular structures and functions. However, the form of the body also determines developmental possibilities. So, it's important to keep in mind that the form, which grounds the essence of the body, not only grounds what will develop but also what *could* develop in various scenarios, including atypical scenarios.

To summarize this section, according to neo-Thomistic hylomorphism, biological regularities across a species are grounded in the same type of form across the species, and the same is true for the regularity of biological possibilities. Members of a species develop standard biological characteristics as a result of having the same type of form. Likewise, the regularities of biological possibilities and impossibilities across a species are grounded by the sameness of form type across the species. The following principle encapsulates the ideas from this section that inform the explanation of NCC that I offer in the next section:

*Forming Biological Regularities* (FBR): The form of the body fundamentally grounds biological regularities in the body throughout a species, such as regularities regarding macro and microscopic organization, structure, and function of the body's constituent biological parts.

FBR finalizes the conceptual framework for the Mind-Body Powers model of NCC. In the section "Aristotelian Powers Ontology," I introduced the first principle, abbreviated IPP, which says there are interdependent partner-powers. Accordingly, things have active and passive powers that co-manifest. Building upon IPP, in the section "Mind-Body Dependence," I introduced the second principle, MBP, which says human persons have what I call mind-body powers, which are mental powers of the soul that are naturally co-manifested with their physical interdependent partner-powers in the body. The section "Mind-Body Powers" provided a further explication of mind-body powers, out of which emerged a principle concerning requisite physical properties, that is, RPP. This principle suggests that biological parts of the human body must have sufficient physical properties to naturally manifest bodily powers that co-constitute mind-body powers. The regularity of standard biological properties in the human species, which are exemplified by parts of the body that manifest the bodily powers of mind-body powers, is determined by the form of the human body. FBR captures this notion, which has been the focus of this section that finalizes the conceptual underpinnings of the proposed model of NCC, to which we now turn.

## MIND-BODY POWERS MODEL

This section presents the Mind-Body Powers model of NCC. The model provides a metaphysical explanation of NCC that's informed by Aristotelian causation as well as neo-Thomistic hylomorphism and fundamentally relies on mind-body powers.

While I am providing an explanation of NCC informed by a dualist human ontology, my account has similarities with non-dualist Aristotelian accounts of human action.[20] Consider, for example, how William Jaworski's hylomorphic view avoids causal overdetermination, the worry that physical causes do all the causal work and irreducible mental causes do not make any real causal contribution. With this problem in mind, Jaworski (2016, p. 281) clarifies the distinct roles of the mental and the physical:

> Beliefs and desires cause or contribute to actions in one kind of way—they rationalize actions—and neural events cause or contribute to actions in a different kind of way—they trigger muscular subsystems involved in actions.[21]

On this account of human action, the mental cause is not reducible to the physical cause, and both are needed. Since the action requires both the mental and the physiological causes, there is no causal overdetermination, as both the mental and the physical make a causal contribution.

It will become apparent that my account of NCC shares common ground with Jaworski's account of human action. However, there are also key differences. First, Jaworski is addressing the problem of causal overdetermination with respect to mental causation, whereas I am providing an account of neural correlates of consciousness. Second, he defends a non-dualist hylomorphic view, whereas my position is a version of substance dualism. That said, Jaworski's account of human action is similar to my account of NCC in the following way: both claim that the mental and the physical make distinct yet interdependent causal contributions.[22] The mind-body powers central to my model of NCC consist of interdependent mental powers and bodily powers manifested by physical parts of the body.

In the following subsection, I will introduce the Mind-Body Powers model of NCC by illustrating it with an example involving a bee sting, in which the mental power is passive. The model will then be further clarified in light of Chalmers's definition of an NCC, and a diagram of the model will be provided. Subsequently, I will offer a second example involving an active mental power in order to further elucidate the model.

## Mind-Body Powers and NCC: Example I

Suppose a man named Freddy is walking barefoot in his yard when he is stung by a bee. To illustrate how mind-body powers apply to NCC, let us consider the sensation Freddy feels when the bee stings his foot.

Freddy's mental state—a conscious sensation of a sharp, stinging pain—is an irreducible mental state that he has first-person access to, which resulted from physical events. When Freddy feels the sting, his mental power to consciously feel such is co-manifested with bodily powers that are manifested by neuronal events in his nervous system. The manifestation of his mental power is the passive result of the manifestation of the bodily powers that the neurophysiological processes in his nervous system are manifesting. He feels the conscious sensation of a vivid sting because a bee stinger penetrated his skin and injected melittin, causing a physiological reaction, which led to countless synapses and action potentials carrying information to his somatosensory cortex.[23] The neuronal states and events manifest bodily powers that produce a change in Freddy's experience, and his conscious mind then manifests its passive power to feel that sharp, stinging sensation, thanks to a little bumble bee and its tiny, yet potent, stinger.

Freddy's mental power is manifested via his subjective conscious state of pain resulting from neurobiological events in his body manifesting the corresponding body powers. The mental and bodily powers co-constitute the mind-body power to feel a bee sting. There is a mutual interdependence between Freddy's mental power (which we can label $MP^{sting}$) manifested via his conscious state and the bodily powers (which we can label $BP^{sting}$) manifested by his neurophysiology. Since the interdependent mental and bodily powers naturally co-manifest the conscious state and neuronal processes manifesting them likewise correspond, resulting in a correlation between the neural mechanisms and the conscious state.

This is why there is a correspondence between conscious states and their neural correlates, according to the Mind-Body Powers model of NCC.[24] The mental power manifested via the conscious state and the bodily powers manifested via the neurobiological states and neurophysiological processes are

158    Chapter 7

interdependent partner-powers, which naturally co-manifest. There is a correspondence between the conscious state and the neuronal substrate by virtue of the metaphysical fact that the mental and bodily powers they manifest are interdependent partner-powers. Furthermore, the bodily powers require sufficient physical properties for their manifestation, according to RPP (see the section "Requisite Physical Properties"). The same type of physical properties in Freddy's nervous system that make it possible to manifest the bodily powers (i.e., $BP^{sting}$), which co-manifest with his mental power (i.e., $MP^{sting}$), will regularly be present in the nervous systems of human organisms throughout the human species, according to FBR (see the section "Biological Regularities").

On this view, human persons have mind-body powers consisting of interdependent mental powers of the soul and body powers manifested by biological body parts. Although these partner-powers are mutually interdependent and co-manifest, they are distinct. Therefore, it is not surprising that the epistemic access to the manifestations of each type of power differ. A person can have direct awareness of the fact that their mental power to consciously think about arithmetic, for example, is being manifested. Yet, they are dependent on empirical studies using neural imaging to know what neuronal events manifest corresponding powers of their body. The corresponding bodily powers require sufficient physical properties, from macro to micro levels of human anatomy, for their manifestation. Due to this, parts of the brain down to the neuronal level with sufficient physical properties will manifest bodily powers that correspond with conscious states manifesting interdependent mental powers. Moreover, the corresponding physical properties will be consistent across the human species given that the same type of form en-forms individual human bodies. Through the tools of cognitive neuroscience, we can learn about what these physical properties are.

## The Model and Chalmers's Definition

Now that I have introduced the Mind-Body Powers model, we can further clarify it by considering it in light of Chalmers's definition of an NCC. After labeling all the various aspects of the mind-body power and the NCC in our above example involving Freddy's bee sting, I will discuss how these aspects map onto the elements of an NCC in Chalmers's definition. Then a diagram will be presented that displays the relationship between NCC and mind-body powers.

When Freddy or any healthy human feels a bee sting, bodily powers that I've labeled $BP^{sting}$ are manifested by neuronal mechanisms, which we can label $N^{sting}$, that have sufficient physical properties to manifest these powers. Action potentials transport information from Freddy's peripheral nervous system into his central nervous system and all the way to his somatosensory cortex, where neuronal mechanisms have physical properties sufficient for manifesting $BP^{sting}$. The manifestation of $BP^{sting}$ via $N^{sting}$ mutually manifests with Freddy's mental power to feel a bee sting, which I've labeled $MP^{sting}$. This mental power is manifested via his conscious state of feeling that painful, sharp, stinging sensation, which we can label $C^{sting}$.

Putting it all together, $C^{sting}$ manifests $MP^{sting}$ which is a partner-power of $BP^{sting}$ manifested via $N^{sting}$. As a result, $N^{sting}$ is a neural correlate of $C^{sting}$. We can now bring to bear on the model Chalmers's (2000, p. 31) definition of an NCC:

> An NCC is a minimal neural system N such that there is a mapping from states of N to states of consciousness, where a given state of N is sufficient, under conditions C, for the corresponding state of consciousness.

In our example involving a bee sting, the *minimal neural system* is $N^{sting}$. And there is a *mapping* from $N^{sting}$ to the *state of consciousness*, $C^{sting}$, because the powers manifested by each ($BP^{sting}$ and $MP^{sting}$, respectively) are interdependent partner-powers, which naturally co-manifest. The *conditions C* are background conditions (e.g., the heart pumping blood to the nervous system) that enable $N^{sting}$ and $C^{sting}$.

## Diagram: Mind-Body Powers and NCC

If, as Aquinas thought, our rational cognition relies on phantasms (or sense images),[25] then there is no substitute for a good diagram that displays ideas to be seen with the eye, insofar as possible. To diagram the model, let's allow the skinny double arrow to represent the metaphysical interdependence between the mental and bodily powers. Let's also allow the wide double ⇔ to represent the resulting natural correlation between the neural mechanisms and the conscious state. In addition, two vertical lines can represent the bodily and mental powers being manifested through the corresponding neural mechanisms and conscious state, respectively. Figure 7.1 can now represent the fundamental ideas in the Mind-Body Powers model of NCC through a diagram.

The bottom level represents the active bodily powers manifested by the neural mechanisms on the left and the passive mental power manifested through the conscious state on the right. These are the interdependent mental and bodily powers constituting the mind-body power to feel a bee sting. The top level represents the correspondence between the neural mechanisms and the conscious state, which is empirically studied through neuroscience. In other words, the NCC is represented on the top level. On the left is the neural correlate manifesting the bodily powers that naturally co-manifests with the mental power manifested via the conscious state on the right.

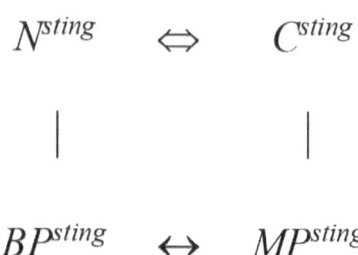

**Figure 7.1  Mind-Body Power to Sense Bee Sting.** Bodily powers $BP^{sting}$ manifested via neural mechanisms $N^{sting}$ with partner mental power $MP^{sting}$ manifested via the conscious state $C^{sting}$.

To illustrate the Mind-Body Powers model, I have used the example of Freddy's bee sting and the associated conscious state along with the corresponding neural correlate. Although Freddy is just a hypothetical healthy adult and the NCC is also hypothetical, if they were real, this diagram would not just reflect Freddy's NCC, but an NCC across the human species.[26] The top level of the diagram would reflect a natural neurobiological regularity across humanity. Therefore, the double arrow on top (that is: ⇔) would be representative of a natural biological regularity, or biological law. However, it is important to note that the arrow does *not* represent a relationship of *necessity*.

According to the definition of NCC provided by Chalmers (2000, p. 31), which has been widely embraced, neural correlates are said to be *sufficient* for the corresponding conscious state, not necessary.[27] An important reason for thinking of neural correlates as sufficient for consciousness, but not necessary, has to do with multiple realizability. Speaking as though a bee actually stung Freddy, the top double arrow ⇔ represents a logically, metaphysically, and physically possible NCC that is an actual NCC in the actual world. Yet, philosophers commonly think it is possible for beings such as animals or even aliens with different physiology than humans to have some of the same conscious states that humans have.[28]

Given multiple realizability, an alien with different "neurons" composed of different materials could be stung by a bee and have the same conscious state as Freddy (i.e., $C^{sting}$). Consequently, it would have a different "neural" correlate of the same conscious state. Assuming the possibility of multiple realizability, a viable account of NCC must be consistent with it and avoid ruling it out. In addition to the more far-out examples of multiple realizability, like the one just mentioned, there are also more mundane examples. For instance, cases where human subjects lose a part of their brain with NCC corresponding to certain types of conscious states, but due to the neuroplasticity of the human brain neural mechanisms in a different part of the brain begin to fulfill the same role as the previous NCC. Loosely speaking, this is like multiple realizability taking place in the same human brain.

In light of such possibilities, it is important to keep in mind that the top double arrow ⇔ is not representing a relationship of logical, metaphysical, nor physical necessity, but rather a relationship of physical sufficiency (see M. Owen and Guta 2019). It represents the biological regularity in the actual world of the conscious state $C^{sting}$ consistently corresponding to the neural correlate $N^{sting}$ in the human brain. Is it possible for a human person to undergo a brain injury and lose $N^{sting}$ and yet have the same conscious state with a different neural correlate, say $N^{sting\text{-}alternative}$? According to the Mind-Body Powers model of NCC, the answer is "yes" as long as $N^{sting\text{-}alternative}$ has physical properties sufficient for manifesting $BP^{sting}$. Since it is the partner-power of $MP^{sting}$ manifested by $C^{sting}$. While such is possible and actually happens in cases where new neural correlates are established via neuroplasticity, there are also many cases of brain damage that result in a permanent loss of natural mental function. In such cases, physical properties sufficient for manifesting the bodily powers that mental powers rely on for their manifestation are not restored.

Let's return to the related issue of whether it's possible for an animal, or even an alien with different physiology and different "neural" correlates than ours (which we can label $N^{sting}*$) to have the same conscious state we have when stung, that is, $C^{sting}$. This is also possible, according to the Mind-Body Powers model, as long as the physical correlates have sufficient physical properties to manifest the requisite bodily powers. In such a case, there would be a different substrate (i.e., $N^{sting}*$) that manifests the bodily powers, $BP^{sting}$, which mutually manifest with the mental power, $MP^{sting}$. Such a scenario can be represented through a new diagram in Figure 7.2.

The difference between this new diagram and the previous one (see Figure 7.1) is that it replaces $N^{sting}$ with $N^{sting}*$, representing the idea that there is a different physical correlate (of the conscious state) that manifests $BP^{sting}$.

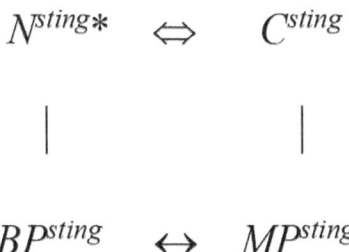

**Figure 7.2 Mind-Body Power to Sense Bee Sting (Alternative System).** Bodily powers $BP^{sting}$ manifested via physical mechanisms $N^{sting}*$ with partner mental power $MP^{sting}$ manifested via the conscious state $C^{sting}$.

According to the Mind-Body Powers model, we do not know what physical states, or, more precisely, what neural states, correlate with particular conscious states. The model is not designed to tell us this. But we do know, assuming the model, that mental powers manifested through conscious states will correspond with body powers manifested by neural mechanisms sufficient for manifesting the body powers. To use our example of the bee sting once more, we know that the natural manifestation of $MP^{sting}$ via $C^{sting}$ requires the co-manifestation of $BP^{sting}$ via some physical mechanisms with the requisite physical characteristics sufficient for manifesting $BP^{sting}$. Accordingly, we have reason to think that there are such physical mechanisms manifesting the body powers if we know that $MP^{sting}$ is being manifested via $C^{sting}$. This provides strong motivation for trying to discover what physical mechanisms in the body manifest $BP^{sting}$. And once we find out what physical mechanisms manifest $BP^{sting}$ and we can discern the manifestation of $BP^{sting}$ by empirically observing the physical mechanisms, we can indirectly discern the co-manifestation of the partner mental power, $MP^{sting}$. The ultimate upshot is that there's good reason to do empirical investigations using the tools of modern neuroscience in search of the physical mechanisms corresponding to consciousness at the neuronal level.[29]

## Mind-Body Powers and NCC: Example II

Since our previous example involving Freddy's bee sting included a passive mental power, I will now briefly provide a second example in which the mental power is active. To begin with, consider Dennis Patterson's act of signing his signature, as described in his review of M.R. Bennett and P.M.S. Hacker's insightful book *The Philosophical Foundations of Neuroscience*:

> Suppose I place my signature on a document. The act of affixing my signature is accompanied by neural firings in my brain. The neural firings do not "explain" what I have done. In signing my name, I might be signing a check, giving an autograph, witnessing a will or signing a death certificate. In each case the neural firing may well be the same. And yet, the meaning of what I have done in affixing my signature is completely different in each case. These differences are "circumstance dependent," not merely the product of my neural firings. Neural firings accompany the act of signing but only the circumstances of my signing, including the intentions to do so, are the significant factors in explaining what I have done. (Patterson 2003)

Patterson's act involves neural processes leading to the movement of his hand scribbling his signature upon a document. Yet the neurophysiological events take place in virtue of Patterson's mental decision to write his signature, which is a constituent of his action.

As we apply the Mind-Body Powers model to Patterson's act, let us suppose that he intended to sign his signature on a document in order to commit to a contract. The requisite neuronal processes in Patterson's brain manifest body powers of biological parts in his body, which co-manifest with Patterson's mental power to intend to sign a contract that is manifested via his conscious intention to sign the contract. In this case, the active power is the mental power manifested through Patterson's conscious intention to sign the contract.

Without the mental power of intending to sign the contract being manifested through his conscious intention, Patterson's act would not be what it is—that is, an act of signing a contract through the physical action of writing his signature. This fits with Aquinas's (ST 1a 80.2c) claim that "a passive capacity takes its proper nature from its relationship to what acts on it." The physical processes resulting in the scribbling of Patterson's signature is what it is—the signing of a contract—in virtue of his conscious intention to sign the contract, which manifests an active mental power. This coheres well with the fact that contracts often have a statement in bold, right above the signature line, that says something like: "I have read this contract completely and fully understand the conditions of this contract."[30] The goal is to verify that the signer consciously intends to sign a binding contract through writing her signature. So, if I somehow connivingly convinced Michael Jordan to sign his autograph on a sheet of paper that was unbeknownst to him a contract giving me all the rights to his shoe sales, the contract would not be legally binding. After all, Jordan would not have truly signed a contract. Since he only intended to sign his autograph for a fan, but not a contract releasing millions of dollars in shoe sales, his action is not an act of signing a contract.

Returning to Patterson's action, Figure 7.3 contains a diagram representing Patterson's act of signing a contract from the perspective of the Mind-Body Powers model of NCC. Patterson's mental power to intend to sign a contract (represented $MP^{sign}$) is manifested via his conscious intention to sign a contract (represented $C^{sign}$) which relies on the co-manifestation of bodily powers (represented $BP^{sign}$) manifested via neural events (represented $N^{sign}$) leading to the movement of his hand scribbling his signature. Each of these elements of the action are portrayed in the diagram.

Like the previous diagrams, the active power is on the left and the passive power is on the right in Figure 7.3. The bottom row represents the mental and bodily powers that are the interdependent partner-powers constituting Patterson's mind-body power to intentionally sign a contract. The mental power on the bottom left is manifested through the conscious state on the top left, and the body powers on the bottom right are manifested through

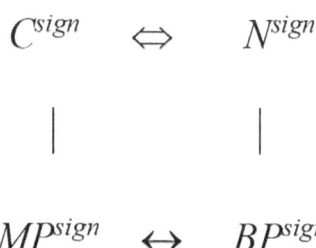

**Figure 7.3 Mind-Body Power to Sign Check.** Mental power $MP^{sign}$ manifested via the conscious intention $C^{sign}$ with partner bodily powers $BP^{sign}$ manifested via neural mechanisms $N^{sign}$.

the neurophysiology on the top right. The top row represents the conscious intention and the corresponding neuronal activity, which is the NCC. What explains the correspondence between the conscious intention and the neural activity is that they manifest interdependent partner-powers which mutually manifest, as co-constituents of a mind-body power.

As we conclude this section in which the Mind-Body Powers model of NCC has been presented, I want to offer an important clarification. As mentioned above, in the section "Aristotelian Powers Ontology," causation involves active and passive powers, but exactly how causation in nature involves these powers is very complex. A single event might manifest a passive power and an active power simultaneously, and multiple active powers might be co-manifested with one passive power or in tandem with a multiplicity of passive powers. There are many examples of such complexity in neurobiology. To offer just one example, consider multiple dendrites of a pyramidal neuron receiving numerous synapses which manifest active powers co-contributing to the manifestation of the neuron's passive power to fire resulting in the manifestation of the neuron's active power to send an action potential along its axon bringing about a synapse at the synaptic cleft where the axon connects with the dendrite of another neuron. As we consider things on this small scale, one begins to see the complexity of causation emerge and the plethora of active and passive roles. Neural processes are simultaneously manifesting both active and passive powers, and there is a multiplicity of active and passive powers co-contributing to particular changes in the neuron. This merely gives us a hint of the causal complexity in the brain, which quickly multiplies as one considers its billions of neurons and the various levels of activity, from neurotransmitters to long-range electrical signals sent from one cortical area to another.

The Mind-Body Powers model of NCC is not intended to represent the causal complexity within the nervous system. It is not a model of *neuron-to-neuron* interaction but a model of *mind-to-body* interdependence. Rather than explaining physical relations in the brain among neurons, it explains a mental-to-physical relation between conscious states and neuronal mechanisms. That's the objective of the model, and it explains neural correlates of consciousness by appealing to the mind-body powers manifested by neural mechanisms in the brain and conscious states of the mind.

## CONCLUSION

Aquinas (ST 1a 77.3c) claimed that the human soul is "at the boundary between spiritual and corporeal creatures, and consequently the powers of both come together in it." With this general idea as a starting point, I have constructed a metaphysical explanation of NCC. To accomplish this objective, principles derived from Aristotle's metaphysics of causation and Aquinas's human ontology have been employed. I hope to have shown that neo-Thomistic hylomorphism is not merely consistent with NCC but also contributes to a good metaphysical explanation of NCC.

The Mind-Body Powers model of NCC explains neural correlates of consciousness by appealing to the interdependence of mental and bodily powers manifested via conscious states and neural mechanisms. Since the mental and bodily powers constituting mind-body powers mutually manifest, their manifestations via neural mechanisms and conscious states correspond. Consequently, there are neural correlates of consciousness. In the final chapter, the Mind-Body Powers model will be applied to the issue of empirically discerning and quantifying consciousness. As it is so applied, the model will be further expounded.[31]

## NOTES

1. See also De Haan (2018), De Haan and Meadows (2014), LaRock (2013), Lowe (2006, 2000), Moreland (2016), and Oderberg (2005).

2. "Hylomorphism" is often spelled "hylemorphism." Oderberg uses the latter spelling.

3. Italics mine.

4. While Lowe is Aristotelian leaning, he does not endorse hylomorphism.

5. According to the problem of overdetermination, irreducible mental causes are moot because physical causes do all the causal work (see chapter 3, section "Causal Exclusion Problem").

6. On the value of such explanations, see Tahko (2012, pp. 37, 39, 40–42), Laudan (1977, Ch. 2), and Lowe (1998, p. 9).

7. Moreland (2018, p. 105) makes a very similar claim. The difference, however, is that he claims his Aristotelian-Thomistic metaphysical framework and human

ontology "even *entails* the search for specific neurological causal/functional/dependency conditions associated with the actualization of the soul's capacities for consciousness" (Moreland 2018, p. 105, italics mine). Interestingly, Kim (2005, p. 124) says that some substance dualist views entail neuronal correlations, and Moreland apparently claims this is true of his view. What exactly is meant by "entail" can make an important difference here and have important implications. Therefore, I do not claim that neo-Thomistic hylomorphism and the Aristotelian metaphysics of causation I adopt in the following pages *entail* NCC. Rather, I am interested in what can provide a *good explanation* of NCC that also helps us better understand their implications. In light of this, my objective aligns with Koch's (2012, p. 121) claim: "Making sense of the discoveries at the mind-brain hinge [i.e., NCC] . . . demands a large-scale, logically consistent framework."

8. Klaus Corcilius (2015, pp. 32–33) makes the point that *dunamis*, according to Aristotle, also applies to the ability to be. Regardless, my account of NCC focuses on powers as the capacity to produce/undergo change.

9. Such practice is common in all areas of study. For example, many of us might be surprised to learn that a neuron's axon can be a meter long because all the example images of neurons we've seen scale the image for teaching purposes. As a result, the axon usually appears to be only about two to three times longer than the width of the cell body. While these images are useful for teaching, we have to keep in mind that things are more complex.

10. Powers are irreducible, according to this view. Contra Hume (2007, p. 55), powers on the Aristotelian view are not a mere projection of ours onto reality, but rather powers are objective features of reality (see Mayr 2011, Chs. 6 and 7). Entities in the world have powers that cannot be reduced to nonpowers. Following Ryle (1949, p. 31), attempts have been made to reduce powers (or dispositions) by giving a conditional analysis of our statements about powers, or by reducing powers to nonpower properties (cf. Choi and Fara 2016; Mayr 2011, Chs. 6 and 7). According to the Aristotelian powers ontology I'm advocating, such reductivist strategies do not sufficiently describe powers, which are a real feature of the world.

11. This idea is seen throughout their works regarding human nature. For an example in Aristotle, see *Metaphysics* (I 1), and for Aquinas, see ST 1a 75.4 ad 1.

12. On the four internal senses, see Pasnau (2002a, p. 281).

13. Italics mine.

14. Quoted from the translation of the *Summa* by the Fathers of the English Dominican province. In this passage Aquinas goes on to say that the intellect and will do not rely on a bodily organ (cf. Aquinas, ST 1a 76.5 ad 2). For an explanation of how this can be consistent with his statements suggesting that rationality relies on the body, see the section "Mind-Body Dependence."

15. For example, to sense we need a sensory system of some *type*. This allows for the possibility that someone who has lost the ability to sense through touch because of damage to their spinal cord could come to exercise their sensory ability via technology that bypasses dependence on the person's spinal cord and somehow sends a signal to the brain. Yet, such an alternative sensory system would need to satisfy the

principle of having sufficient physical properties to manifest "bodily" dispositions involved in sensing.

16. See chapter 2, specifically section "What is an NCC?."

17. See the *Allen Brain Map* at brain-map.org.

18. Italics mine.

19. See also Marmodoro and Page (2016, p. 16). They also emphasize that a substantial form is not an efficient cause.

20. It may also share commonality with views not labeled Aristotelian. For example, perhaps there is some common ground with Chinook Won's (2019) defense of the idea that physical effects can be jointly caused by mental properties and their physical bases.

21. Cf. Donald Davidson (1963).

22. On this point, as alluded to above, I think we share common ground with Haldane (1998, pp. 271–272), Moreland (2016, p. 117), and Oderberg (2005, p. 90).

23. For an introduction to the somatic sensory system, see Bear et al. (2016, Ch. 12).

24. Cf. Moreland (2016, p. 117) and Oderberg (2005, p. 90).

25. See the section "Mind-Body Dependence."

26. While we can speak of homogeneity and sameness regarding neurobiology and NCC across a species, it is important to remember that we are speaking of homogeneity regarding biological regularities that permit some degree of variation. Variations pertaining to NCC can be due to variations in overall conscious experiences or variations in individual brains across a species (see chapter 2, section "What is an NCC?").

27. Koch's (2019b, p. 48) definition follows suit and likewise defines NCC as sufficient not necessary (see also Koch et al. 2016, p. 307; Mormann and Koch 2007). It is best to understand the sufficiency of NCC according to physical sufficiency, as opposed to logical or metaphysical sufficiency (M. Owen and Guta 2019). On the wide embrace and influence of Chalmers's definition of NCC, see M. Owen and Guta (2019, p. 2).

28. For prominent examples, see Saul Kripke (1981) and Hilary Putnam (1967).

29. Cf. Moreland (2018, p. 105).

30. At least this is often the case in my home country, the United States.

31. A prior iteration of this chapter was published in a special issue of *Topoi* entitled "Mental Powers," edited by Matteo Grasso and Anna Marmodoro. Reprinted by permission from Springer Nature: Springer Nature, Topoi, (Owen, Matthew, "Aristotelian Causation and Neural Correlates of Consciousness"), Copyright © 2020. https://doi.org/10.1007/s11245-018-9606-9

*Chapter 8*

# Empirically Discerning and Measuring Consciousness

The question of whether we could empirically detect the presence of consciousness and measure someone's level of consciousness sounds like an abstract topic. Yet for the neurologist trying to treat an unresponsive patient with a severe brain injury, the ability to empirically assess whether the patient is conscious and to what degree would be an invaluable diagnostic tool (see the section "Diagnosing Disorders of Consciousness"). However, the possibility of being able to empirically detect and measure consciousness without requiring any response from the patient is inherently linked to how consciousness relates to the brain (Gosseries et al. 2014, pp. 457, 466, 468; Boly et al. 2013, p. 2).

If consciousness were identical to or determined by a neurophysiological process, it would be theoretically possible to empirically discern and measure consciousness. But if, contrary to physicalism, consciousness is irreducible and does not supervene on neurobiology, then the possibility of empirically discerning and quantifying consciousness might seem devoid of any realistic foundation. In this chapter, I will argue to the contrary that the Mind-Body Powers model of NCC provides a framework for understanding how consciousness relates to the brain that grounds the possibility of empirically discerning and measuring irreducible, nonphysical consciousness.[1] This will show that there is a nonphysicalist framework that undergirds the objective of empirically detecting and measuring consciousness, which is critical to the science of consciousness.

In making my case, I will combine the Mind-Body Powers model with a hypothesis about the neural correlate of being conscious provided by the Integrated Information Theory of consciousness. I will do so in order to demonstrate the metaphysical possibility of empirically detecting and quantifying irreducible, nonphysical consciousness in unresponsive patients. Given this possibility, physicalists and nonphysicalists alike have warrant for trying to make this theoretical possibility a practical actuality at the bedside of patients who are unresponsive yet potentially conscious.

Before focusing on the theoretical basis for empirically discerning and measuring consciousness, I will begin this chapter by clarifying what "consciousness" means in this context. Then, in the section "Diagnosing Disorders of Consciousness," I discuss the importance of accurately diagnosing disorders of consciousness and why doing so can be difficult. The goal is to clarify why it would be medically advantageous to be able to empirically discern whether a patient is conscious and to what degree. Subsequently, in the section "Indirectly Measuring Consciousness," I explain what it would mean to measure, or quantify, irreducible consciousness and the role of NCC. In the section "The Full NCC," two hypotheses about the neural correlate of being conscious, which is called the *full NCC*, are introduced and contrasted before I subsequently focus on one of them. As already alluded to, the hypothesis I focus on is provided by the Integrated Information Theory (for brevity IIT). Further empirical research is needed to confirm or disconfirm what IIT predicts about the neural activity correlated with being conscious. Here I simply use IIT's prediction as a plausible and perspicuous example of what the neural correlate might be.

In the section "Mind-Body Powers and the Full NCC," the Mind-Body Powers model is combined with IIT's hypothesis about the full NCC to demonstrate how the model provides a metaphysical basis for the possibility of empirically discerning and quantifying irreducible consciousness, even when patients are completely unresponsive. Before concluding, I consider the implications of artificially stimulated neural activity in postmortem brains and offer caveats about adjudicating death.

## WHAT IS CONSCIOUSNESS?

In the following section, various disorders of consciousness (abbreviated DOC) are discussed along with the challenges of accurately diagnosing such disorders. As indicated by the phrase "disorders of consciousness," the conditions that distinguish the different disorders pertain to consciousness. Therefore, it is important to be as clear as possible about what we mean by *consciousness* when discussing DOC.

It is often said that the best things in life are free. Though it is less commonly noted, they can also be difficult to define. Love, freedom, and humor provide a few examples. Even something as fundamental to our world as physical space is difficult to define without appealing to physical space in the definition itself. While consciousness is integral to our lives, or at least our waking lives, it resists a succinct definition that applies to every instance of it (see Chalmers 1996, pp. 3–11). Sometimes we are consciously aware of an object in our visual field, or a musical melody at a concert. At other times, we sit with our eyes closed in a quiet, still library and consciously reflect on whether a series of premises logically entails a particular conclusion. Each morning we emerge from a sleepy slumber. Although we are conscious in our first waking moments, we may not be consciously aware of anything in particular.

There can be various scenarios where consciousness is present even though the accompanying mental capacities differ. Consciousness might be accompanied by rational cognition, visual awareness, emotional moods, desires, or mental intentions, and this does not exhaust the possibilities with respect to the accompanying capacities. Yet, what is integral to consciousness itself is *experience* (Chalmers 1996, p. 4; Koch 2019b, p. 1). What I fundamentally mean by "consciousness" is experience. If a subject is conscious, they have an experience. I am concerned in this chapter with whether it is possible to empirically discern whether an unresponsive patient is experiencing and how vivid their experience is. Consciousness is often accompanied by rational thought, attention focused on an object, or language use for the sake of communication. However, given that experience is most fundamental to consciousness it is important to know whether patients are experiencing, even if they are not exercising other mental capacities.

## DIAGNOSING DISORDERS OF CONSCIOUSNESS

There are various disorders of consciousness (for brevity DOC) that have different conditions associated with them. For our purposes, there are three relevant DOC.

The first is coma. This is a pathological state where the patient is unconscious and there is a continuous absence of eye-opening, and behaviors are limited to only reflexive movements (Schnakers et al. 2013, p. 118). The vegetative state (abbreviated VS) is the second relevant DOC. This disorder is also referred to as unresponsive wakefulness syndrome (abbreviated UWS), which is more fitting since these patients open their eyes spontaneously or in response to stimulation, but their behaviors are only reflexive and unrelated to their environment (Schnakers et al. 2013, p. 118; Laureys et al. 2010). The third DOC, the minimally conscious state (abbreviated MCS), is often distinguished by the presence of inconsistent but clear, recognizable behavioral signs of consciousness (Schnakers et al. 2013, p. 119).[2]

Accurately diagnosing disorders of consciousness is crucial for various reasons. To begin with, the patient's prognosis can correspond to their diagnosis. For example, there is reason to believe a diagnosis of MCS corresponds with a better prognosis with a more favorable chance of functional recovery than a diagnosis of VS/UWS (Giacino 2004, p. 296; Giacino and Kalmar 1997). Secondly, a patient's diagnosis affects plans for how to most effectively treat the patient and care for them, which affects their quality of life (see Giacino et al. 2014). For example, if it's known that a patient has emerged from an unconscious state and is now minimally conscious, pain management becomes more important. Moreover, different medical departments implement different rehabilitation programs for patients based on their particular disorder of consciousness. A patient's diagnosis can often determine which department they are admitted to and therefore the rehabilitation support they receive (Giacino et al. 2014, p. 103). There are also standards insurance companies have for medical expenses to be financially covered that are based on the patient's diagnosis (Giacino et al. 2014, p. 103).

In addition to the aforementioned practical and clinical reasons, there are serious ethical reasons why accurately diagnosing DOC patients is important (see Giacino et al. 2014, pp. 108–109; Fins et al. 2007; Fins 2015). Whether or not a patient is conscious can correspond to their likelihood of clinical improvement (see Owen 2020, p. 1052). Therefore, it is important to know whether a patient is conscious when family members and healthcare providers are deciding whether or not to continue life support. Misdiagnosing a patient who is behaviorally unresponsive but nevertheless covertly conscious can lead to a premature withdrawal of life-sustaining treatment (Giacino et al. 2014, p. 103). The legality of withdrawing life support is also affected by the patient's diagnosis. In several nations, withdrawing artificial hydration and nutrition is legal in VS/UWS cases but not in MCS cases (Schnakers et al. 2016, p. 178; see Manning 2012).

While accurately diagnosing DOC patients is clinically and ethically important, it can be very difficult. Since discerning whether a patient is conscious and to what degree comes with various epistemic challenges discussed below. A study conducted by Caroline Schnakers and her colleagues is sobering. They found that 41 percent of forty-four patients in their study who were diagnosed as VS/UWS patients were actually MCS (Schnakers et al. 2009, 2016, p. 171). More recently, a study led by Jiahui Pan implemented a novel machine learning algorithm in the assessment of DOC patients (Pan et al. 2020). In a scientific commentary on the study, Adrian Owen points out that 40 percent of the patients in the study clinically diagnosed as VS/UWS patients were able to follow commands well enough to be reclassified as aware (Owen 2020, p. 1052). Reports regularly suggest that roughly 30–40 percent of patients diagnosed as VS/UWS are actually conscious, and thus misdiagnoses are far more common than we can be comfortable with (Giacino et al. 2014, p. 103; cf. Massimini and Tononi 2018,

p. 33). Next I will clarify the complexities of accurately diagnosing DOC patients. This will elucidate why it would be useful to have a way to empirically discern the presence of consciousness that does not require patients to give any response.

## Behavioral Response

Behaviors that indicate consciousness, which are often called behavioral correlates of consciousness, are commonly used to discern the degree to which a patient is conscious. Medical practitioners regularly observe behavioral responses to provided stimuli in order to assess a patient's level of consciousness (see Giacino et al. 2004; Teasdale and Jennett 1974). Whereas an articulate verbal response to a complex question indicates a higher level of consciousness, a reflexive response to a sensory stimulus can indicate a lower level. While behavioral correlates of consciousness can be useful, their utility has prerequisites, limiting their effectiveness.[3] One prerequisite is that patients must be conscious enough to perceive the external stimuli and respond appropriately. Yet a patient could be conscious but nevertheless not conscious enough to perceive an external stimulus and respond appropriately.

A second prerequisite is that the patient's sensory systems must be functioning well enough for them to sense an external stimulus. However, someone could be conscious while their sensory systems fail to function properly. An obvious example is when someone's vision is completely impaired—they do not become unconscious simply because they cannot see. Similarly, a patient who has undergone a severe brain injury might have damaged sensory systems, compromising their ability to perceive external stimuli. Even if such a patient is conscious, she might remain behaviorally unresponsive to external stimuli not because she is unconscious but because she cannot sense the stimuli.

A third prerequisite is that the patient's motor systems must be sufficiently preserved to carry out a behavioral response that necessitates motor movement. However, a patient could lack motor movement while retaining consciousness. For example, a patient with locked-in syndrome (LIS) might be considered unconscious because they lack the motor movement to respond adequately to a command or a sensory stimulus. Nevertheless, this does not mean they are unconscious, as we know from LIS patients who have minimal motor movements preserved that allow them to communicate using eye blinks (Massimini and Tononi 2018, p. 32). Such communication makes it evident that the patient is conscious despite lacking the motor movement required to give standard behavioral responses.[4]

## Neuronal Response

Another paradigm for discerning the presence of consciousness relies on neuronal responses from the patient, which can be used to discern the presence of consciousness in patients lacking motor movement required for a behavioral response. This paradigm was first used by a team of neuroscientists led by Adrian Owen and Steven Laureys at the University of Cambridge and the University of Liege. They showed that a brain-injured patient lacking motor movement yet retaining consciousness could provide a neuronal response to imaginative tasks (Owen et al. 2006).

Laying the foundation for this neuronal response paradigm, Mélanie Boly and her colleagues did a study on healthy subjects and found that when they followed imaginative commands there was specific neural activity corresponding to their specific mental activity (Boly et al. 2007). When told to imagine playing tennis, there was corresponding neural activity in the supplementary motor area, which is involved in planning movement. Additionally, when they were told to imagine walking through their home, different coalitions of neurons that correspond to processing spatial coordinates and memory of locations were activated in the posterior parietal cortex, the parahippocampal gyrus, and the lateral premotor cortex (see Boly et al. 2007; Owen et al. 2006; Massimini and Tononi 2018, p. 37).

Having mapped the distinct neural activity that corresponds to each mental activity, the team was able to use this information to empirically discern the presence of consciousness when a severely brain-injured patient lacked motor movement. As was done with healthy subjects, the researchers could give the commands and observe the corresponding neural activity one would expect if, in response to the command, the patient imagined playing tennis and at another time imagined walking through their home (Owen et al. 2006, p. 1402).[5] This paradigm is helpful for detecting covert consciousness in behaviorally unresponsive patients, and neuronal responses have also been used as a way for patients to provide simple "yes/no" answers to questions (Owen 2020, p. 1051; Monti et al. 2010).

The neuronal response paradigm is remarkable and useful for diagnosing patients who retain the ability to sense the external world but completely lack motor movement. That said, it still requires an active response from the patient. Rather than requiring a behavioral response, it requires a neuronal response. There are various reasons why patients who are conscious might nevertheless be unable to provide even a neuronal response (Boly et al. 2013, p. 3).

Many severely brain-injured patients are aphasic and thus lack the ability to comprehend the language in a command such as "imagine playing tennis." Other patients may be unable to give a neuronal response because the region of their brain that would be activated once they understood the command and tried to respond is damaged (Massimini and Tononi 2018, p. 39). Then there are some patients who may just be too depressed, unmotivated, or confused to respond to what seems to them like a pointless game (Massimini and Tononi 2018, p. 39). Furthermore, just like a behavioral response, a neuronal response requires patients to have sensory systems that are functioning adequately enough for them to sense a stimulus in their environment. Specifically, they must be able to sense the command, whether it be a command to imagine playing tennis or to imagine navigating through their home. Overall, the neuronal response paradigm still requires patients to be able to sense the external world and to be conscious and cognizant enough to respond appropriately (Giacino et al. 2014, p. 105).

**Passive Paradigm**

For the aforementioned reasons, the so-called active paradigms that require a behavioral or neuronal response from the patient are too insensitive to detect when a completely unresponsive patient is nevertheless conscious (see Boly et al. 2013, p. 3; Giacino et al. 2014, p. 105; Gosseries et al. 2014, p. 462). While a response is indicative of consciousness, a lack of response is uninterpretable since the lack of a response does not indicate the absence of consciousness (Gosseries et al. 2014, p. 462). What is needed is a passive paradigm that allows us to detect and measure consciousness without requiring the patient to give any response at all (Gosseries et al. 2014, p. 463). "Indeed, in the absence of a gold standard for measuring consciousness, active functional neuroimaging or neurophysiological tools must strike a balance between avoidance of false-positive and false-negative errors" (Giacino et al. 2014, p. 106). To accurately diagnose completely unresponsive DOC patients, we need a means for discerning the presence and level of consciousness that relies on an empirically recognizable and measurable indicator of consciousness. Simply put, we need to be able to empirically detect and quantify consciousness.

However, one might reply: you don't always get what you want, nor even what you need, despite how helpful it might be. That's true. Nonetheless it is also true that sometimes you do get what you want and what you need, but only after toiling toward an objective that is difficult to reach but perhaps possible. As mentioned earlier, physicalism provides a metaphysical foundation for the theoretical possibility of empirically discerning and quantifying consciousness (see chapter 1). My goal is to show that there is a nonphysicalist framework that does likewise. I intend to demonstrate in this chapter that the Mind-Body Powers model grounds the metaphysical possibility of empirically recognizing and quantifying irreducible, nonphysical consciousness. Having now clarified the medical motivation behind this objective, let's consider what it would mean to measure consciousness if it is not reducible to its neural mechanisms.

## INDIRECTLY MEASURING CONSCIOUSNESS

When it comes to quantifying consciousness, it is important to distinguish between direct and indirect measurements. Direct measurements (e.g., using a tape measure to measure the length of a wall) often come to mind when we think about measurements. Yet, in practice, indirect measurements are very common.

Imagine that you have arrived at Paradise Lodge on Mount Rainier on a beautiful August day. Since the weather is conducive for a long hike, you begin flirting with the idea of climbing up to Camp Muir, the base camp for climbers attempting to summit the following day. But you do not have any overnight equipment, so you want to be back to Paradise by nightfall. To find out whether it is possible to make it to Muir and back to Paradise in time, you need to measure the distance from Paradise to Muir. There are multiple ways to do so. One way is to measure the distance directly using a surveyor's wheel (i.e., a hodometer). As you walk up the trail from Paradise to Muir, the total number of revolutions the wheel makes will tell you how many meters you have covered. However, a much easier way to measure the distance is to use a map of Mount Rainier that includes points representing Paradise Lodge and Camp Muir. While Mount Rainier is not reducible to the image on the map, the space on the map between the points representing Paradise and Muir is a reliable indicator of the actual distance between the two real physical locations. This is because there is a correlation between the space on the map between the two points and the distance between the two places. So after measuring the space on the map, you can then infer the distance between Paradise and Camp Muir. By directly counting the centimeters between the two points on the map, you are indirectly measuring the actual distance between the two locations, which is roughly 7 kilometers.

For a more mundane example, consider time. As Augustine (*Confessions*, 11.14.17) quipped, we know precisely what it is, as long as no one asks us what it is. If anything is abstract and irreducible, time is a clear contender. Yet we measure it, albeit indirectly. Every full rotation of the second hand around the clock corresponds to one minute. Every rotation of the minute hand corresponds to an hour. One need not be a philosopher of time to figure out that time is not reducible to these measurements. By measuring the movements of the second hand, we are not directly measuring time itself. Rather, we are indirectly measuring time by directly measuring reliable correlations—that is, the movements of the second hand.

In psychological studies, we might measure levels of stress by measuring cortisol levels (see, e.g., Hunter et al. 2019). However, I highly doubt the stress I feel about an upcoming presentation can be comprehensively described by a mere reference to cortisol. Yet, stress need not be reducible to a hormone in order for a measurement of the hormone to be indicative of stress levels. As long as there is a consistent correlation between the rise of cortisol levels and stress levels, then a measurement of the former can be used to indirectly measure the latter.

These are just a handful of examples of indirect measurements that rely on correlations. I assume that irreducible consciousness is not directly measurable, but that does not mean it cannot be indirectly measured based on reliable correlations (cf. Chalmers 1998, pp. 220–221). As discussed above, purposeful behaviors in response to stimuli or commands often provide reliable behavioral correlates of consciousness. Yet when behavioral and even neuronal responses are absent, another correlation is needed if we are to reliably discern whether consciousness is present and to what extent without requiring a response from the patient.

This is where neural correlates of consciousness (NCC) can become imperative. Recall from chapter 2 that in NCC research an important distinction is made between the *full NCC* and *content-specific NCC* (see Koch et al. 2016, p. 308; Storm et al. 2017). Content-specific NCC are the minimal neural mechanisms that, together, are physically sufficient for a particular state of consciousness with specific content, such as seeing Nikki Haley's face on your iPhone screen or hearing the horn from the car behind you. The full NCC are the minimal neural mechanisms that, together, are physically sufficient for being conscious or one's overall conscious experience generally regardless of specific content. The full NCC is most relevant to empirically discerning and quantifying consciousness in unresponsive patients.

## THE FULL NCC

The nature and location of the neural correlate minimally sufficient for consciousness to be present is currently unknown. While various content-specific NCC have been identified, the full NCC is yet to be discovered. Two predictions about the nature and location of the full NCC are provided by the Global Neuronal Workspace theory (for brevity GNW) and the Integrated Information Theory (abbreviated IIT).[6] GNW is a computational theory of consciousness, whereas IIT is not (see Dehaene et al. 2017, p. 492; Koch 2019b). In this section, I will introduce both theories but then focus specifically on IIT's prediction about the full NCC as an example of a plausible prediction about the neural correlate's nature and location. I do not take a stance on which theory is best or whether there is an alternative theory that is more promising. Rather, to make my overall point regarding empirically discerning and quantifying consciousness, I rely on IIT's prediction about the full NCC simply as an example of a reasonable hypothesis.

Chronologically speaking, the Global Neuronal Workspace theory is first in line. The theoretical neurobiologist, Bernard Baars, proposed GNW before the turn of the century, soon after which the neuroscientist and psychiatrist, Giulio Tononi, first published on IIT (see Baars 1988; Tononi 2004). Baars posited that within the brain there is a global workspace that stores information and makes it available to the specialized systems throughout the brain.[7] Given that the workspace's capacity is limited, information signals compete to be the globally available representation in the workspace. Whatever perception or thought that comes to occupy the workspace is conscious as it is projected to various systems throughout the brain. The percepts or thoughts that do not make it into the workspace, and therefore are not globally available, are not conscious.

Contemporary advocates of the Global Neuronal Workspace theory have applied it to the neurophysiology of the neocortex (see Dehaene and Changeux 2011; Dehaene 2014). Accordingly, a marker of consciousness is a global broadcast of information involving the activity of a prefrontal-parietal network of long-range cortical neurons corresponding with activity in high-level sensory cortices receiving the broadcast. Such activity indicates that the information is globally available for various functional processes such as speech, memory, or action and thus conscious content, per GNW. If, however, there are no long-range projections and neuronal activity is very localized in specific areas, that would indicate there is no global broadcast of information making it available to specialized systems throughout the brain. It would follow that consciousness is not present, given GNW.

As we will see, IIT makes a very different prediction about the nature and location of the full NCC and therefore suggests a different marker of consciousness. Although the first difference worth mentioning pertains to IIT's method of inquiry. IIT does not claim to start with the brain and then consider how consciousness emerges from it. Rather, IIT is forthrightly based on five starting self-evident axioms about the nature of consciousness from which five postulates are inferred about the nature of the physical substrate of consciousness in the brain (Tononi et al. 2016, p. 450; Tononi and Koch 2015, p. 5).[8]

The first axiom of IIT, called *intrinsic existence*, says consciousness exists and is intrinsic to the subject of the conscious experience who has direct epistemic access to it.[9] Based on this axiom it is postulated that the physical substrate of consciousness (abbreviated PSC) must exist and produce intrinsic causal effects upon itself. The second axiom, *composition*, says conscious experience is structured in that it has distinguishing features. Given this, IIT postulates that the constituent elements of the PSC must themselves, or, together with other elements, have causal power upon the system. The axiom of *information* says that consciousness is specific in that each experience is distinguished from other conscious experiences due to its distinct phenomenological features. It is postulated from this that the PSC exhibits a cause-effect structure of a specific form that makes it distinct from other possible structures. According to the fourth axiom, *integration*, a conscious experience is a unified whole irreducible to the phenomenal distinctions within it. Thus, IIT postulates the cause-effect structure exemplified by the PSC must be unitary and irreducible to noninterdependent causal subsystems. The final axiom, *exclusion*, claims a conscious experience is definite in content, spatial perception, and temporal duration. Therefore, IIT postulates the cause-effect structure exemplified by the PSC will also be definite, including no more and no less than a set of elements exhibiting causal power on the whole.

According to IIT, the PSC exemplifies a structure in the central nervous system that exhibits maximal intrinsic cause-effect power (Tononi et al. 2016, p. 450). This maximally irreducible cause-effect structure is the integrated information that is consciousness, according to the theory (see Koch 2019b, pp. 87–89). The integrated information is referred to as *Phi* and represented $\Phi$, with the vertical bar seen as representing information and the circle integration (Massimini and Tononi 2018, p. 72). Yet, it is the maximal Phi, represented $\Phi^{max}$, in a system that is said to be consciousness. While there might be pockets of neural activity throughout the brain manifesting positive $\Phi$, it is that which is manifesting the maximal $\Phi$, or the highest degree of integrated information, that's said to be the subject's overall conscious experience.

The axioms and postulates of IIT mentioned above concern human consciousness and the human nervous system. However, the theory is often extrapolated and applied to consciousness in other biological organisms with a nervous system similar to humans. And the theory is applied even more broadly to all conceivable manifestations of consciousness, with the idea being that a positive $\Phi$ in any physical system is indicative of consciousness. Elsewhere, I have offered constructive criticism of this universally broad application of the theory (see M. Owen 2019b, section 5). Here I am not concerned with what the theory predicts about consciousness in other biological organisms and physical systems. Rather, I am interested in what it predicts about human consciousness and the PSC in the human nervous system. And to be even more precise, I am only concerned with what the theory predicts about the PSC of being conscious, not the PSC of specific conscious states such as various states of qualia. In other words, what matters to my present aim is what IIT predicts about the full NCC in humans and the nature of human consciousness. This is what's relevant to empirically discerning and quantifying consciousness in unresponsive human patients.

Given that consciousness is an intrinsic causal structure in the central nervous system, the causal structure would indicate consciousness is present, and by measuring it one could measure consciousness. Put differently, if a measurement of brain activity at the neuronal level in an unresponsive patient produced a positive Φ measurement, then we could empirically know consciousness is present and to what degree. Since this would be a measurement of the causal structure consciousness is identical to. However, this process depends on consciousness being identical to the causal structure in the PSC that the Φ measurement is a measurement of. But there is reason to doubt this identity relation, which I will discuss in the following section. Without the identity relation, however, one needs a metaphysical framework that provides a basis for inferring the presence and level of consciousness in an unresponsive patient from the Φ measurement. As I will go on to demonstrate, the Mind-Body Powers model yields such a framework.[10]

## Consciousness and Phi

Hedda Hassel Mørch (2019a) has recently given a metaphysically motivated reason to doubt that consciousness is identical to maximal integrated information, or $\Phi^{max}$, manifested by the PSC.[11] She appeals to the intrinsicality problem, which highlights an apparent inconsistency between consciousness and maximal Phi. According to this line of argument, maximal Phi and consciousness cannot be identical since consciousness is intrinsic, whereas maximal Phi is extrinsic. Maximal Phi is extrinsic in that it depends on the properties of something else, namely the amount of Phi manifested elsewhere in the brain (or whatever physical system is being considered). When there are two manifestations of Phi present in the central nervous system—$\Phi^1$ and $\Phi^2$—whether $\Phi^1$ is maximal will depend on a property of $\Phi^2$, namely its Phi level, and vice versa. But consciousness is not external in this way and therefore not identical to maximal Phi. Given this difference, Mørch (2019a, pp. 136, 156–159) recommends rejecting the idea that consciousness is identical to or reducible to $\Phi^{max}$. Mørch's argument appeals to an ontological difference between maximal Phi and consciousness. Yet IIT's epistemic methodology also provides grounds for an epistemologically motivated objection to the idea that consciousness is reducible to maximal Phi.

IIT's epistemic methodology starts with self-evident axioms about the nature of consciousness that we can be directly aware of. From these axioms, postulates are inferred about the nature of the physical substrate of consciousness, which is said to manifest intrinsic causation or an intrinsic causal structure. There is an important epistemic difference between the axioms about consciousness and the postulates about the PSC manifesting intrinsic causation. The axioms regarding the nature of consciousness are self-evident, directly accessible to the conscious person, whether or not one has the brain imaging technology required to observe neural activity or any causation at all in the brain. However, the nature of the PSC manifesting the intrinsic causal structure in the brain is not self-evident, nor directly knowable, but rather empirically known (assuming the postulates are accurate). Hence, IIT's postulates need to be tested and confirmed or falsified via empirical neuroscientific a posteriori investigation.

Overall, IIT's methodology suggests that the nature of consciousness is self-evident and directly knowable,[12] whereas the causal structure manifested by the PSC is not. The nature of consciousness can be known by a conscious person without observing the brain. On the other hand, knowledge of the causal structure manifested by the PSC requires observation of the brain. Thus, it seems most reasonable to conclude that consciousness is not identical to the causal structure the PSC exemplifies.

To drive the point home further, we can apply the rendition of Jackson's (1982, p. 130) thought experiment presented in chapter 2, which pertains to direct epistemic access to one's conscious states.[13] However, this time let's consider Mary's sister, Kary, who has been deaf since birth. Like Mary, Kary is also a neuroscientist, and despite the fact that she has never heard their songs, she, too, happens to be a major fan of the *Red Hot Chili Peppers* because she likes the meaning of their lyrics. Kary studies the neural activity corresponding to conscious experiences that are the result of drug use. And she has developed a red pill that causes people to fall asleep and dream that they are at the Hollywood Bowl in an entirely red world listening to the Red Hot Chili Peppers play their hit songs backward. Kary has designed a study to find out what the physical substrate of this far-out experience is. Since they had a positive experience participating in Mary's experiment and they're curious about what their songs sound like backward, the Red Hot Chili Peppers band members decide to participate as subjects in Kary's study, as they did in Mary's.

So Kary gives them the red pill and observes their brain activity as they experience the strange dream. She finds out that the experience corresponds to neural activity manifesting intrinsic causation in the occipital and temporal lobe with a very high Phi measurement. In the end, she learns everything there is to know about this causal structure manifested in the PSC. However, she keeps this information entirely to herself and does not share it with the band. For she remembers that when Mary did her study, due to a falling out with the lead singer, the drummer left the band to pursue a career in neuroscience. Concerned another band member might do likewise and try to publish the outcome of the study before her, like Mary, Kary also keeps all the data to herself.

However, despite her comprehensive knowledge of the causal structure in the PSC that corresponds to the strange dream, Kary has never had the conscious experience herself. Therefore, she does not know what-it's-like to dream the strange dream nor what-it's-like to hear the hit songs played backward. As a result, the band members have knowledge Kary lacks. She knows all about the intrinsic causal structure in the PSC, but the band members know what-it's-like to consciously experience the dream and to hear their songs played backward. Kary, on the other hand, does not know what-it's-like since the way to get epistemic access to the conscious experience itself is to experience it, which she has never done. Nevertheless, she knows all about the intrinsic causation manifested by the PSC of the experience, which the band members know nothing about. For such knowledge requires empirically observing the brain, which they have never done.

The epistemic contrast suggests the conscious experience and the intrinsic causation correlated with it are distinct, and not identical. The conscious experience is directly accessible to the experiencer, whereas the causal structure is not. Moreover, the causal structure can be known by someone not having the conscious experience and completely unknown to the experiencer. Consequently, the conscious experience and the corresponding intrinsic causation differ.

However, if consciousness is not identical to the intrinsic causal structure that we could theoretically measure, this raises a problem for inferring the presence and level of consciousness based on the presence and measurement of the causal structure. As mentioned above, if consciousness is identical to $\Phi$, then it would follow that consciousness is present given that $\Phi$ is present. Although if they are not identical, you need further information that confirms a correspondence between consciousness and the intrinsic causal structure. This is where reports from subjects are very useful. But when subjects are completely unresponsive yet perhaps covertly conscious, reports are unavailable. Yet, further information is needed nevertheless, if we are to reliably infer that consciousness is present on the basis of a positive $\Phi$ measurement in the PSC. At this point, the Mind-Body Powers model can help by providing additional information from which inferences concerning the presence of consciousness can be made. The model can contribute to an overall theoretical framework that could aid interpretations of empirically observable neural activity in the brain of an unresponsive patient.[14]

## MIND-BODY POWERS AND THE FULL NCC

The previous chapter discussed how the Mind-Body Powers model of NCC provides a metaphysical explanation of neural correlates of consciousness.

Here I will summarize the model and its strengths before explicating its application to empirically discerning and quantifying irreducible consciousness in unresponsive patients. To show how it applies, I will combine the model with IIT's prediction about the full NCC itself—namely, that consciousness corresponds to a maximally irreducible cause-effect structure in the PSC—while assuming that consciousness is not identical to the NCC.

The metaphysics of Aristotelian causation that informs the Mind-Body Powers model is very important to its applicability to discerning and measuring consciousness. Recall that according to Aristotelian causation, there are causal powers that "depend for their activation on the activation of their mutual partner-powers" (Marmodoro 2014a, p. 32). Thus, according to an Aristotelian powers ontology, there are interdependent partner-powers that are naturally co-manifested. As Marmodoro (2014a, p. 32) puts it: "Because of their mutual dependence for activation, partner-powers realize their natures in activities that are *co-determined*, *co-varying*, and *co-extensive* in time."

According to the neo-Thomistic hylomorphic human ontology informing the model, a person is a single material substance naturally consisting of a soul that's the substantial form of the body. The soul is a powerful entity, and due to human nature it relies on the body to manifest its powers (see Aquinas, ST 1a 76.5c). Like God and angels, the nature of human persons is rational but it is also sensory, like animals, according to Aquinas (see SCG II.46, 57; ST 1a 89.1c and 75.7 ad 3). Given this, and that he followed Aristotle's thought that the body is required for the soul to sense, Aquinas considered the body necessary for human persons to operate consistently with their nature and therefore essential to human nature (see Aquinas, SCG II.57, ST 1a 84.4c, 89.1c, 75.4c, 75.7 ad 3).[15] He thought the body has the variety of organs it has so the soul can manifest its various powers or capacities (Aquinas, ST 1a 78.3c, 78.4c). Different capacities rely on different organs that have different structures sufficient to manifest different powers (see Aquinas, QDA 8c, ad 16). And as this is true of all organs, it is equally true of the brain (see Aquinas, QDA 8c).

Aquinas thought about the soul's dependence on the body to exercise its powers in light of the scientific understanding of his time. Hence, he thought about these issues at the level of organs, not the cellular level that was unobservable during his day, which predated the first use of the term "neuron" and Santiago Ramón y Cajal's neuron doctrine by six centuries (cf. Finger 2000, Ch. 13). The Mind-Body Powers model of NCC applies Aquinas's thought in light of what we now know from neurobiology to the cellular level. According to the model, human persons have both mental powers of the mind and bodily powers of biological structures in the body that are interdependent partner-powers, which I call mind-body powers. The power to feel what-it's-like to taste Swiss chocolate is a mental power. But this power relies on and is naturally co-manifested with bodily powers manifested by biological parts in the body that are suitably structured, from the macro level of the tongue to the micro level of neurons in the gustatory cortex.

The power to experience what-it's-like to see the Matterhorn's north face is also a mental power, but it's not the same mental power as the capacity to feel what-it's-like to taste Swiss chocolate. Consequently, it co-manifests with a different set of bodily powers manifested by biological body parts which are sufficiently structured—from the organic level of the eyes, to brain structures such as the optic chiasm, to neurons in the visual cortex. Since the corresponding bodily powers are distinct, they are manifested by distinct biological body parts with particular structures sufficient to manifest the particular body powers.

Thus, as different organs correspond to particular mental powers, which Aquinas recognized, different structures in the brain even at the neuronal level correspond to particular mental powers. This is because the mental powers rely on the manifestation of bodily powers in biological parts that are suitably structured. The correspondence between consciousness and neuronal mechanisms is grounded in the metaphysics of mind-body powers. For a conscious state and its corresponding neural activity correlate with consistent regularity because they manifest a mental power and bodily power that are interdependent partner-powers that naturally co-manifest, as co-constituents of a mind-body power. And since the mental and bodily powers naturally co-manifest as co-constituents of a mind-body power, the manifestation of one can indicate the manifestation of the other. Hence, it is not surprising that once a researcher knows what the NCC of seeing Bill Clinton's face is, the researcher can know when a subject sees Clinton's face by observing the NCC via brain imaging (see Koch 2012, pp. 65–66).

The Mind-Body Powers model of NCC does not tell us what neural states and processes correlate with particular conscious states. Rather, it leaves this to neuroscience research aimed at identifying what specific neuronal structures, states, and processes are physically sufficient to support particular conscious states. What it does, however, is provide a metaphysical explanation of the correspondence between conscious states and their neural correlates that warrants the search for these physically sufficient neuronal mechanisms of consciousness. After all, the bodily powers, which are the interdependent partner-powers of the mental powers manifested through states of consciousness, require sufficient physical properties in the body to be manifested. Success in the search for NCC tells us what these physical properties are.

In addition to providing a good metaphysical explanation of neural correlates of consciousness, the model has multiple benefits worth briefly mentioning (even though I cannot give a full explanation of each one here). First, the model is consistent with the irreducibility of consciousness. The model does not explain the correspondence between consciousness and NCC by identifying the former with the latter. Therefore, one is not left with the epistemic burden of finding a way to counterintuitively explain how consciousness can be identical to physical brain processes in the face of powerful arguments to the contrary.[16] Yet, at the same time, the model is also consistent with the intuition that our bodies are essential to human nature. For the model is based on Aquinas's idea that the soul relies on the body to operate consistently with its human nature and is consequently integral to human nature. Moreover, the Thomistic human ontology informing the model can account for the unity of consciousness, the simplicity of the self, and the persistence of an individual human person through time and bodily change.[17] According to neo-Thomistic hylomorphism, the ultimate bearer of consciousness is a metaphysically simple soul not composed of parts that persists through time as the body it en-forms significantly changes through growth and aging. While these strengths of the Mind-Body Powers model are worth noting, here I want to focus on a strength of the model with a potential practical payoff in medical contexts where brain-injured patients are completely unresponsive.

The model can provide grounds for the metaphysical possibility of empirically discerning and quantifying consciousness in unresponsive patients even though consciousness is not reducible to a physical entity that we can directly measure. To demonstrate this, let us look to IIT's hypothesis about the full NCC in the human nervous system as an example of a clear prediction about what the NCC might be. As discussed above, according to IIT's prediction, the presence of consciousness will correspond with intrinsic causation manifested by a coalition of neurons. Leading proponents of IIT predict that such integrated neural activity constituting the full NCC is likely in the posterior cerebral cortex localized in a temporo-parietal-occipital "hot zone" (Koch et al. 2016, pp. 308 and 315). In other words, the full NCC is likely a coalition of neurons sufficiently structured for reciprocal projections in sensory areas toward the back of the brain. Verification of this will require further empirical testing that is currently underway.[18] For our purposes here, I am using IIT's prediction merely as an example of what the full NCC might be in order to demonstrate the applicability of the Mind-Body Powers model. So for the sake of demonstrating the metaphysical possibility of empirically discerning and quantifying irreducible consciousness, let us here assume that IIT's prediction is accurate and has been validated.

To elucidate how mind-body powers can ground this possibility given IIT's prediction about the full NCC, we can employ the diagram illustration of the Mind-Body Powers model from the previous chapter (see Figure 7.1). The full NCC, as described by IIT, is represented by $\Phi$ and plugged into the diagram illustration in Figure 8.1.

The metaphysical interdependence of powers that explains the correspondence between a specific conscious state and a content-specific NCC likewise

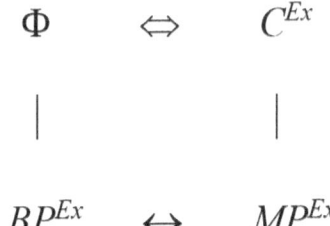

**Figure 8.1  Mind-Body Powers Model and the Full NCC.** Bodily power $BP^{Ex}$ manifested, via the full NCC represented $\Phi$, with partner mental power $MP^{Ex}$ manifested via the conscious state $C^{Ex}$.

undergirds the correspondence between the overall state of being conscious and its corresponding full NCC. The mental and bodily powers the mental state and neural activity manifest are interdependent partner-powers. On the right is the mental power to consciously experience (represented $MP^{Ex}$) manifested through the conscious state of experiencing (represented $C^{Ex}$). On the left is the interdependent partner bodily power (represented $BP^{Ex}$) manifested by intrinsic causation among a coalition of neurons (represented $\Phi$) in the thalamocortical system, which support the manifestation of the mental power to consciously experience. Together the mental and bodily powers are interdependent partner-powers constituting the mind-body power to be conscious.

Given that they are interdependent partner-powers that together constitute the mind-body power to be conscious, the mental and bodily powers naturally co-manifest with each other. Consequently, if the mental power manifests, then the bodily power manifests; and if the bodily power manifests, then the mental power manifests. So, assuming IIT's prediction is empirically proven veridical, the presence of intrinsic causation in the coalition of neurons that's the full NCC would strongly indicate the presence of consciousness. Because the intrinsic causation among the neurons is a manifestation of a bodily power (or powers)[19] that naturally co-manifests with the mental power to consciously experience, since they are interdependent partner-powers.

Thus, when IIT's prediction about the full NCC is coupled with the Mind-Body Powers model, we are given an empirically discernible basis for recognizing the presence of consciousness. By measuring the intrinsic causation among the coalition of neurons that's the full NCC (plausibly in a temporo-parietal-occipital hot zone), we would be measuring the manifestation of a bodily power that directly corresponds to the mental power to consciously experience. Granted, we would not be directly measuring consciousness itself given that it is irreducible to the physical substrate that is being measured (see the sections "Indirectly Measuring Consciousness" and "Consciousness and Phi"). Nevertheless we would be indirectly measuring consciousness by directly measuring its neural correlate that corresponds to consciousness itself and the degree to which someone is conscious.

The practical reliability of this method of inquiry has at least two clear prerequisites. The first is the confirmed legitimacy of the Mind-Body Powers model of NCC. To its credit, the model appears to have the strengths mentioned above, but further work focused on substantiating this is in order. The second prerequisite is empirical confirmation of the full NCC's location and nature.

Here I have relied on IIT's prediction as an example of a clear prediction about the nature and location of the full NCC. Yet all that the model needs for discerning the presence of consciousness is the identity of the full NCC, whether or not it is what IIT predicts. For given the model, the presence of the full NCC is strongly indicative of the presence of consciousness. The benefit of IIT's prediction, however, is that if it is true, it would provide not only an empirically discernible "footprint" of minimal consciousness but also a way to measure minimal levels of consciousness. For it is possible that a patient be somewhere along a spectrum of minimal consciousness, either closer to or further away from being capable of responding to stimuli while they are presently unresponsive. Given IIT's prediction about the full NCC, the amount of intrinsic causation corresponds to the level of consciousness—the greater the causation, the greater the level of consciousness.

Thus far, I have argued that in light of the Mind-Body Powers model the presence of consciousness is strongly indicated by the manifestation of a bodily power(s) via neural activity which is the full NCC. This is because the bodily power of the physical substrate and the mental power to consciously experience naturally co-manifest as co-constituents of the mind-body power humans have to be conscious. With this conclusion in mind, a recent study on postmortem pig brains prompts a relevant question that can help clarify my claim.

## ARTIFICIAL STIMULATION

In 2019, a team of researchers at the Yale School of Medicine partly revived postmortem pig brains (see Vrselja et al. 2019; Farahany et al. 2019; Reardon 2019). Some might be under the impression that the study showed it's possible to artificially stimulate neural activity in postmortem brains. In reality, the scientists used synaptic blockers to preclude neural activity and consequently the brains had an entirely flat EEG.[20] Nevertheless, the study elicits a germane hypothetical question: If neural activity that's naturally the full NCC in a human brain were artificially stimulated postmortem, would the activity indicate consciousness from the perspective of the Mind-Body Powers model?

There are at least two significant interrelated factors relevant to the answer. One factor is whether the experiencer—that is, the person—is present or if there is reason to think the person is absent. A second factor is what is meant by "*artificial* stimulation."

With respect to the first factor, as Descartes thought thinking requires a thinker—I think therefore I (the one thinking) am—experience requires an experiencer. Consciousness is not a free-floating property devoid of a subject but is rather a state of a subject, the bearer of consciousness, which is in this context the human person (Guta 2019b, p. 132). Consequently, if the person is not present, consciousness is not present. In light of the Aristotelian-Thomistic thought informing the Mind-Body Powers model, once the body is no longer alive, the soul has ceased exercising its rational, sensitive, and vegetative capacities.[21] And the body, on this view, is no longer en-formed by the human soul that once en-formed it. Therefore, the person would not be present with the body, which is more precisely called a corpse, not a body. Even if it were to undergo substantial change in which it became en-formed by a different substantial form, that form would not be a human soul.[22] From this perspective, there would be no metaphysical reason to think the person remained present. And even if the person were present with the dead body, once the brain is removed from the substantial organism of which it was an inseparable part, it would take on a different form. From an Aristotelian-Thomistic perspective, I do not see any reason to think the new form the brain would then be en-formed by would be a human soul or another kind of form with the capacity to be conscious.

One might contest, however, that since neural activity is sufficient for consciousness, the stimulation of such activity logically entails the presence of consciousness, regardless of whether the person is present or not. But that is only true if the neural activity is logically sufficient for consciousness. As I have argued at length elsewhere, with Mihretu Guta, NCC are best understood as physically sufficient for consciousness, not logically sufficient (see M. Owen and Guta 2019). Given that, one must also consider whether the metaphysical prerequisites, such as the person being present, are satisfied so consciousness could be present. Moreover, we have no empirical study demonstrating that neural activity devoid of a subject is sufficient for consciousness since all localization studies involve living subjects capable of experiencing. Put differently, in studies identifying NCC in the human brain the person is unquestionably present. To boot, in all the studies where we have been able to map a correspondence between conscious states and neural activity, the brain was still part of a living body. This is significantly different from mapping a correspondence between conscious states and neural activity in a brain removed from a dead body. The fact that consciousness corresponds with neural activity in a brain that's part of a living body simply does not entail that such activity would correspond to consciousness in a brain removed from a corpse.

With that said, let us assume the metaphysical prerequisites were also met for consciousness to potentially be present. In other words, let's assume there is a good reason to think a person would be present with the brain removed from the dead body. Assuming that, we must think carefully about the artificial stimulation of the neural activity.

On the one hand, there is artificially supporting the *background conditions* physically required for the neural processes that are the full NCC to adequately function. An example would be artificially providing cardiovascular circulation necessary but not sufficient for the neural activity. If the brain were removed from a corpse, this type of artificial stimulation would look very different. Yet, regardless of how it was accomplished, it would involve artificially providing the brain with the conditions necessary but not sufficient for it to function. If this were accomplished, then the neural activity that is the full NCC would need an explanation given that the artificial stimulation provided was necessary but not sufficient for the activity. In such a case, the Mind-Body Powers model would provide an answer for why the bodily power is manifested via the neural activity that's the NCC. The answer would be that the bodily power is being manifested because its interdependent partner-power—that is, the mental power to consciously experience—is being manifested.

On the other hand, there is artificially supporting the necessary background conditions *plus* artificially stimulating the neural activity that is the NCC. For example, this might be done by perturbing the neural mechanisms using direct electrical stimulation. If the full NCC were artificially stimulated in this sense, the Mind-Body Powers model would *not* provide justification for thinking consciousness is present given the artificially stimulated neural activity. After all, the neural activity that's not so artificially stimulated indicates consciousness because what explains why the activity manifesting the bodily power is present is the co-manifestation of its partner mental power to consciously experience. If the neural activity itself were artificially stimulated, there would be an alternative explanation of the activity that does not require the manifestation of the partner mental power. The obvious explanation is that the neuronal activity was artificially stimulated. Consequently, the activity would not indicate the presence of consciousness from the standpoint of the model.[23] Since the artificial stimulation of the neural activity would be sufficient to explain the activity, and the manifestation of the mental power to consciously experience is not needed to explain it. It is neural activity that's not artificially stimulated that suggests the corresponding mental power is being manifested with its interdependent partner bodily power.

## CAVEATS

I have focused here on the possibility of empirically discerning and quantifying consciousness in unresponsive patients. My intention has been to demonstrate that it is metaphysically possible given the framework provided by the Mind-Body Powers model combined with IIT's prediction about the full NCC. A related topic is how the presence or absence of consciousness is related to death and epistemically determining when death has occurred. While this topic is very important, it's quite complicated in our modern world with technology capable of sustaining vital functions, such as oxygenation and cardiovascular circulation.[24] Due to the complexity of the topic, I cannot adequately address it here. Nevertheless, I will offer several points of clarification.

I take it for granted that if there is neural activity indicating that the person is conscious, then the person is alive, not dead. Given that, demonstrating consciousness is present is one way to show that the person is alive. It does not follow, however, that if consciousness is absent, then the person is dead. My focus on discerning the presence of consciousness should not be misinterpreted as suggesting that consciousness provides the boundary between life and death. To me, that is obviously false. I have lost consciousness numerous times, every night when I go to sleep, when under anesthesia, in skiing accidents, and so on—yet here I am, living, breathing, writing. It seems quite clear to me that while the presence of consciousness entails life, a loss of consciousness does not entail death. Therefore, we cannot infer someone has died solely on the basis that consciousness is absent. That would be a significant error.

Furthermore, we should note an important distinction between being conscious and being capable of being conscious. When I am asleep, I may not be conscious, but that does not entail that I am not capable of being conscious. (Indeed, my two-month-old daughter seems to know that I am capable of being ushered into a state of consciousness by the sound of crying.) The distinction between being conscious and the capacity to be conscious is important because a patient might be unconscious yet still capable of being conscious. Once again, we can make a positive inference—from the fact that someone is conscious, we can know they are still capable of being conscious. However, we cannot infer that if one is not conscious then they are incapable of being conscious. From the Thomistic perspective I have presented, while the neural activity that is the full NCC can indicate the manifestation of the power (or capacity) to consciously experience, the absence of the activity does not entail the absence of the capacity. For a capacity, or power, can be present yet unmanifested. Thus, as Jason Eberl (2006, p. 48) points out, "while Aquinas notes that one can determine the presence of a certain capacity based upon observation of its corresponding activity (ST, Ia.87.1; Pasnau,

2002[c], pp. 336–341), it does not follow that failure to observe an activity entails the lack of its corresponding capacity."

Yet, even if the capacity for consciousness itself (not just its manifestation) is lost, there is more to consider from an Aristotelian-Thomistic perspective.[25] Because the human soul is a substantial form that is not only rational and sensitive but also nutritive, the manifestation of any of the soul's capacities could indicate its presence and that the person is alive. Accordingly, even when a person does not have access to their cognitive faculties and will not reawake mentally, if their body is still a living organism then the soul would seem to be present, manifesting its nutritive capacity. When exactly the body ceases to be a unified living biological organism can be very difficult to discern. But until it ceases to be such, from the Thomistic perspective, there is reason to think death has not occurred (cf. Eberl 2006, Ch. 3). In sum, while discerning the presence of consciousness can confirm that death has not occurred, it is important to remember that in its absence there is more to consider in light of the neo-Thomistic human ontology I'm advocating.

## CONCLUSION

Reconsiderations of dualism (see chapter 1) might be cause for concern if nonphysical minds are necessarily causally impotent and inconsistent with neurobiology, or inconsonant with empirically discerning and measuring consciousness. However, I have argued that such worries are unwarranted with respect to the dualist view, neo-Thomistic hylomorphism. In chapter 6, I applied the view to mental causation and argued that it provides a coherent account of causal pairing. In chapter 7, I applied it to neurobiology and developed the Mind-Body Powers model of NCC, demonstrating that hylomorphism is not only consistent with neural correlates of consciousness but also informs a good metaphysical explanation of them. In this final chapter, I applied the Mind-Body Powers model to the concern that a nonphysicalist dualist understanding of human nature undermines the possibility of empirically discerning and measuring consciousness, which could be critical for diagnosing disorders of consciousness in unresponsive patients.

I have argued that empirically discerning and quantifying irreducible consciousness is metaphysically possible given the Mind-Body Powers model of NCC informed by neo-Thomistic hylomorphism and Aristotelian causation. Toward this end, I relied on IIT's prediction about the full NCC as a plausible example of the neural correlate's nature. From the standpoint of the Mind-Body Powers model, one can reliably infer that consciousness is present from the presence of neural activity that is the full NCC. This is because the full NCC manifests a bodily power(s) that naturally co-manifests with its interdependent partner mental power to consciously experience. For they are co-constituents of the mind-body power human persons have to be conscious. Moreover, if the full NCC does in fact consist of measurable reciprocal causation among a coalition of neurons that reliably corresponds with being more or less conscious, as IIT predicts, then it would be possible to indirectly measure a subject's level of consciousness. In light of the model, the mental power to consciously experience would be manifested to a corresponding degree that the bodily power is manifested via the neural activity that is the full NCC.

In the final analysis, empirically discerning and measuring consciousness seems possible even from a nonphysicalist dualist vantage point. The Mind-Body Powers model of NCC provides grounds for the metaphysical possibility of empirically detecting the presence and level of irreducible consciousness. Hence, whether this theoretical possibility can become a practical actuality remains an open question even as dualism is dusted off, reconsidered, and potentially undergoes a revival. What remains to be seen, however, is whether nature is so orchestrated that the music of the mind is sufficiently synced with neural mechanisms adequate for recording its melody. Whether nature will reveal that such is so is a mystery physicalists and dualists alike can seek to solve. Perhaps, the "immeasurable" mind will turn out to be measurable after all in the light of future neurobiology. Let *us*—physicalists and dualists alike—try to find out.

## NOTES

1. I use "detect/discern" synonymously, as well as "measure/quantify."
2. There is also a distinction made between MCS- and MCS+ (Gosseries et al. 2014, p. 459; Boly et al. 2013, p. 3).
3. I focus here on three general limitations. Schnakers et al. (2016, p. 171) list specific concerns.
4. Hence, LIS is not technically a DOC (Giacino et al. 2014, pp. 100–101).
5. Cf. Pan et al. (2020); Naci et al. (2014).
6. The global brain signal is another, more recently proposed, candidate marker of consciousness well worth attention (Tanabe et al. 2020). For ease of exposition, I focus here on just two well-known theories. For additional prominent neuroscientific theories of consciousness proposed over the last two decades, see Lamme (2006), Block (2011), and Lau and Rosenthal (2011).
7. For a lucid outline of GNW that informed my own, see Koch (2019b, pp. 139–140).
8. See also Fallon (2016, section 1).
9. My summary of IIT's axioms and postulates is informed by Tononi et al. (2016, pp. 450–452), Tononi (2017, pp. 243–248), and Tononi and Koch (2015, pp. 6–7).

10. Assuming, for the sake of argument, that IIT's prediction about the full NCC is verified by forthcoming research.

11. Elsewhere, I have distinguished between what I call *reductive IIT*, *nonreductive IIT*, and *nonphysicalist IIT*, which differ based on how the ontology of consciousness is understood (Owen 2020d). This section presents critiques of reductive IIT, to which nonreductive IIT and nonphysicalist IIT are immune. While those who wish to fit IIT into a reductionist framework might be attracted to reductive IIT, I don't think it is most consistent with how the leading proponents of IIT understand the theory.

12. More precisely, I understand IIT proponents as claiming that a conscious subject has knowledge by acquaintance of *de re* facts about their consciousness—such as, my conscious experience exists and it's integrated (see Koch 2019b, p. 11; Tononi 2017, p. 243; cf. M. Owen 2020c, p. 149).

13. See section "The Identity Theory's Viability."

14. Regarding the need for and utility of a theoretical framework, see M. Owen and Guta (2019, pp. 9–11).

15. On how this is compatible with statements Aquinas makes about rationality being an independent operation not needing the body and the Christian doctrine of the intermediate state Aquinas held, see chapter 7 section "Mind-Body Dependence" and M. Owen (2020a, section 3). Briefly, Aquinas thought the body is required for the human soul, which is both rational and sensory, to naturally exercise rational cognition. Nevertheless, he thought it is metaphysically possible for the soul to rationally cognize in an alternative way apart from the body (see Aquinas, ST 1a 89.1c, 75.7 ad 3, and QDA 1). Ross Inman has pointed out to me the potential applicability of multitrack dispositions here. One might explain Aquinas's view by appealing to multitrack dispositions and argue that mental powers are naturally co-manifested with bodily powers but capable of co-manifesting with other nonbodily powers in the intermediate state.

16. For examples of relevant arguments, see Chalmers (1996), Mørch (2019b), and Robinson (2016).

17. Regarding relevant discussion of the unity consciousness, see Hasker (2010). Regarding the simplicity of the self, see Lowe (2001) and Barnett (2010). Concerning the persistence of an individual through time and bodily change, see Brown (2005, Ch. 5).

18. For a news release about such research being done through an adversarial collaboration between proponents of IIT and GNW, see Ball (2019).

19. The careful reader will notice that I predominantly use "bodily power" (singular) here when discussing the full NCC, whereas I used "bodily powers" (plural) when discussing content-specific NCC in the previous chapter. When discussing the full NCC as IIT understands it, I think it's fitting to use "bodily/body power" (singular) in reference to that which is manifested by the coalition of neurons exhibiting $\Phi$. Yet this is not meant to suggest that the full NCC only involves a single neurobiological capacity. See chapter 7, section "Mind-Body Powers and NCC: Example II" for an important clarification regarding the causal complexity in the nervous system.

20. I am indebted here to Christof Koch for his comments on an earlier manuscript draft in November 2020. What is most interesting about the study is that the

researchers excited some cortical neurons in neural tissue removed from the postmortem brains using an electrode, demonstrating that some individual neurons retained their capacity for synaptic activity (see Koch 2019a, p. 37).

21. Here I am speaking specifically of the natural exercise of these capacities according to the nature of humanity, the cessation of which does not rule out the soul exercising rationality in an alternative way in the intermediate state apart from the body (see chapter 7, section "Mind-Body Dependence" and M. Owen 2020a, section 3).

22. Aristotelians have, throughout history, debated whether a corpse is en-formed by a different substantial form or is an aggregate of substances with their own substantial forms. I think the latter is correct.

23. One might also wonder if it is even the same activity as that constituting the full NCC, even if it were the same neurons performing indistinguishable functions. Consider an analogous example where I sign a contract because of my intention to commit to its conditions versus my nervous system being artificially stimulated so my hand moves in a way that brings about my signature on the contract (see chapter 7, section "Mind-Body Powers and NCC: Example II" for the inspiration of this example). The former is arguably not the same as the latter even if the physical processes are indistinguishable.

24. See Laureys (2005), Nair-Collins and Miller (2017), Verheijde et al. (2018), Shewmon (1997), Eberl (2006, Ch. 3), Bernat (1998), and Esmaeilzadeh et al. (2010).

25. Cf. Eberl (2006, Ch. 3), Norkowski (2018), Shewmon (1997), and Moreland and Wallace (1995).

# References

Alkire, M.T., Hudentz, A.G., and Tononi, G. 2008. Consciousness and anesthesia. *Science*, 322(5903): 876–880. doi: 10.1126/science.1149213

Allen Institute for Brain Science. 2020. *Allen Brain Map*. Seattle, WA. http://brain-map.org/

Alter, T., and Howell, R.J. (eds). 2011. *Consciousness and the Mind-Body Problem*. New York, NY: Oxford University Press.

Anscombe, G.E.M. (ed). 1981. *Metaphysics and the Philosophy of Mind*. Oxford, UK: Basil Blackwell Publisher.

Aquinas, T. *Commentary on Peter Lombard's Sentences*.

Aquinas, T. 1947. *Summa Theologica*. Translated by Fathers of the English Dominican Province. Benziger Bros. Edition. Christian Classics Ethereal Library.

Aquinas, T. 1949. *Disputed Questions on Spiritual Creatures*. Translated by Fitzpatrick, M.C. and Wellmuth, J.J. Milwaukee, WI: Marquette University Press.

Aquinas, T. 1956. *Summa Contra Gentiles, Book Two: Creation*. Translated by Anderson, J.F. 1975 Edition. Notre Dame, IN: University of Notre Dame Press.

Aquinas, T. 1984. *Questions on the Soul*. Translated by Robb, J.H. Milwaukee, WI: Marquette University Press.

Aquinas, T. 2002. *The Treatise on Human Nature: Summa Theologiae 1a 75-89*. Translated by Pasnau, R. Indianapolis, IN: Hackett.

Aquinas, T. 2012. *The Power of God*. Translated by Regan, R.J. Oxford Scholarship Online. doi: 10.1093/acprof:osobl/9780199914395.001.0001

Aristotle. 1984a. "Metaphysics." In Barnes, J. (ed.), *The Complete Works of Aristotle*, 1552-1728. Princeton, NJ: Princeton University Press.

Aristotle. 1984b. "On the soul." In Barnes, J. (ed.), *The Complete Works of Aristotle*, 641-692. Princeton, NJ: Princeton University Press.

Aristotle. 1984c. "Physics." In Barnes, J. (ed.), *The Complete Works of Aristotle*, 315-446. Princeton, NJ: Princeton University Press.

Audi, P. 2012. Grounding: Toward a theory of the in-virtue-of relation. *The Journal of Philosophy,* CIX (12): 685–711.

Augustine. 1991a. *Confessions*. Translated by Chadwick, H. New York, NY: Oxford University Press.
Augustine. 1991b. *The Trinity*. Translated by Hill, E. Brooklyn, NY: New City Press.
Baars, B.J. 1988. *A Cognitive Theory of Consciousness*. New York, NY: Cambridge University Press.
Baars, B.J. 2003. Introduction: Treating consciousness as a variable: The Fading Taboo. In Baars, B.J., Banks, W.P., and Newman, J.B. (eds.), *Essential Sources in the Scientific Study of Consciousness*, 1–10. Cambridge, MA: MIT Press.
Bailey, A. 2009. Zombies and epiphenomenalism. *Dialogue*, 48: 129–144.
Bailey, A.M., Rasmussen, J., and Horn, L.V. 2011. No pairing problem. *Philosophical Studies*, 154(3): 349–360.
Baker, M.C., and Goetz, S. (eds). 2011. *The Soul Hypothesis: Investigations into the Existence of the Soul*. New York, NY: Continuum.
Ball, P. 2019. Neuroscience Readies for a Showdown Over Consciousness Ideas. *Quanta Magazine*. https://www.quantamagazine.org/neuroscience-readies-for-a-showdown-over-consciousness-ideas-20190306/
Barnett, D. 2010. "You are simple." In Koons, R. C., and Bealer, G. (eds.), *The Waning of Materialism*, 161–174. New York, NY: Oxford University Press.
Bayne, T. 2018. On the axiomatic foundations of the integrated information theory of consciousness. *Neuroscience of Consciousness*, 4(1): niy007. doi: 10/1093/nc/niy2007
Bear, M.F., Connors, B.W., and Paradiso, M.A. 2016. *Neuroscience: Exploring the Brain*. Philadelphia, PA: Wolters Kluwer.
Beebee, H., Effingham, N., and Goff, P. 2011. *Metaphysics: The Key Concepts*. New York, NY: Routledge.
Bennett, M.R., and Hacker, P.M.S. 2003. *Philosophical Foundations of Neuroscience*. Oxford, UK: Blackwell Publishing.
Bernat, J.L. 1998. A defense of the whole-brain concept of death. *The Hastings Center Report*, 28(2): 14–23.
Bhaduri, A., Andrews, M.G., Leon, W.M., Jung, D., Shin, D., Allen, D., Jung, D., Schmunk, G., Haeussler, M., Slama, J., Pollen, A.A., Nowakowski, T.J., and Kriegstein, A.R. 2020. Cell stress in cortical organoids impairs molecular subtype specification. *Nature*, 578: 142–148. doi: 10.1038/s41586-020-1692-0
Bickle, J., Mandik, P., and Landreth, A. 2010. "The philosophy of neuroscience." In Zalta, E.N. (ed.), *The Stanford Encyclopedia of Philosophy*. Summer 2012 Edition. https://plato.stanford.edu/entries/neuroscience/
Blake, R., and Logothetis, N.K. 2002. Visual competition. *Nature Reviews Neuroscience*, 3: 13–21. doi: 10.1038/nrn701
Blake, R., and Tong, F. 2008. Binocular rivalry. *Scholarpedia: The Peer-Reviewed Open-Access Encyclopedia*, 3(12): 1578. doi: 10.4249/scholarpedia.1578
Bliss, R., and Trogdon, K. 2016. "Metaphysical grounding." In Zalta, E.N. (ed.), *The Stanford Encyclopedia of Philosophy*. Spring 2016 Edition. https://plato.stanford.edu/entries/grounding/
Block, N. 2011. Perceptual consciousness overflows cognitive access. *Trends in Cognitive Sciences*, 15(12): 567–575. doi: 10.1016/j.tics.2011.11.001

Block, N., and Stalnaker, R. 1999. Analysis, dualism, and the explanatory gap. *The Philosophical Review*, 108(1): 1–46.
Bogardus, T. 2013. Undefeated dualism. *Philosophical Studies*, 165(2): 445–467.
Bohm, D. 1952. A suggested interpretation of the quantum theory in terms of "hidden" variables. I & II. *Physical Review*, 85(2): 166–193.
Bohm, D., and Hiley, B.J. 1993. *The Undivided Universe: An Ontological Interpretation of Quantum Theory*. New York, NY: Routledge.
Boly, M., Coleman, M., Davis, M., Hampshire, A., Bor, D., Moonen, G., Maquet, P.A., Pickard, J.D., Laureys, S., and Owen, A.M. 2007. When thoughts become action: An fMRI paradigm to study volitional brain activity in non-communicative brain injured patients. *NeuroImage*, 36(3): 979–992.
Boly, M., Massimini, M., Tsuchiya, N., Postle, B.R., Koch, C., and Tononi, G. 2017. Are the neural correlates of consciousness in the front or in the back of the cerebral cortex? Clinical and neuroimaging evidence. *The Journal of Neuroscience*, 37(40): 9603–9613. doi: 10.1523/JNEUROSCI.3218-16.2017
Boly, M., Seth, A., Wilke, M., Ingmundson, P., Baars, B., Laureys, S., Edelman, D.B., and Tsuchiya, N. 2013. Consciousness in humans and non-human animals: recent advances and future directions. *Frontiers in Psychology*, 4 (625). doi: 10.3389/fpsyg.2013.00625
BonJour, L. 2010. "Against materialism." In Koons, R. C., and Bealer, G. (eds.), *The Waning of Materialism*, 3–24. New York, NY: Oxford University Press.
BonJour, L. 2013. What is it like to be a human (instead of a bat)? *American Philosophical Quarterly*, 50(4): 373–385.
Bricmont, J. 2016. *Making Sense of Quantum Mechanics*. Heidelberg: Springer.
Broad, C.D. 1955. Kant's mathematical antimonies: The presidential address. *Proceedings of the Aristotelian Society*, 55: 1–22.
Brown, C.M. 2005. *Aquinas and the Ship of Theseus: Solving Puzzles about Material Objects*. New York, NY: Continuum.
Bunge, M. 1961. The weight of simplicity in the construction and assaying of scientific theories. *Philosophy of Science*, 28 (2): 120–149.
Cattell, J.M. 1893. Mental measurement. *The Philosophical Review*, 2(3): 316–332.
Chalmers, D.J. 1996. *The Conscious Mind: In Search of a Fundamental Theory*. New York, NY: Oxford University Press.
Chalmers, D.J. 1998. "On the search for the neural correlate of consciousness." In Hameroff, S.R., Kaszniak, A.W., and Scott, A.C. (eds.), *Toward a Science of Consciousness II: The Second Tucson Discussions and Debates*, 219–229. Cambridge, MA: MIT Press.
Chalmers, D.J. 2000. "What is a neural correlate of consciousness?" In Metzinger, T. (ed.), *Neural Correlates of Consciousness*, 17–40. Cambridge, MA: MIT Press.
Chalmers, D.J. 2003. "Consciousness and its place in nature." In Stich, S.P. and Warfield, T.A. (eds.), *The Blackwell Guide to Philosophy of Mind*, 102–142. Malden, MA: Blackwell.
Chisholm, R.M. 1976. *Person and Object: A Metaphysical Study*. Chicago, IL: Open Court.
Chisholm, R.M. 1991. On the simplicity of the soul. *Philosophical Perspectives*, 5: 167–181.

Choi, S., and Fara, M. 2016. "Dispositions." In Zalta, E.N. (ed.), *The Stanford Encyclopedia of Philosophy*. Spring 2016 Edition. https://plato.stanford.edu/entries/dispositions/

Chudnoff, E. 2011. What should a theory of knowledge do? *dialectica*, 65(4): 561–579.

Churchland, P.M. 1985. Reduction, qualia, and the direct introspection of brain states. *The Journal of Philosophy*, 82(1): 8–28.

Cooper, J.W. 1989. *Body, Soul & Life Everlasting: Biblical Anthropology and the Monism-Dualism Debate*. Grand Rapids, MI: Eerdmans.

Cooper, J.W. 2009. The current body-soul debate: A case for dualistic holism. *Southern Baptist Journal of Theology*, 13(2): 32–50.

Copan, P., and Craig, W.L. 2018a. *The Kalām Cosmological Argument, Volume 1: Philosophical Arguments for the Finitude of the Past*. New York, NY: Bloomsbury.

Copan, P., and Craig, W.L. 2018b. *The Kalām Cosmological Argument, Volume 2: Scientific Evidence for the Beginning of the Universe*. New York, NY: Bloomsbury.

Copan, P., and Taliaferro, C. 2019. *The Naturalness of Belief: New Essays on Theism's Rationality*. Lanham, MD: Lexington Books/Rowman & Littlefield.

Corcilius, K. 2015. "Faculties in ancient philosophy." In Perler, D. (ed.), *The Faculties: A History*, 20–58. New York, NY: Oxford University Press.

Correia, F., and Schnieder, B. 2012. "Grounding: An opinionated introduction." In Correia, F., and Schnieder, B. (eds.), *Metaphysical Grounding*, 1–36. Cambridge, UK: Cambridge University Press.

Craig, W.L. 1979. *The Kalām Cosmological Argument*. Palgrave Macmillan.

Craig, W.L., and Moreland, J.P. 2009. *The Blackwell Companion to Natural Theology*. Oxford, UK: Wiley-Blackwell.

Cramer, S.C., Sur, M., Dobkin, B.H., O'Brien, C., Sanger, T.D., Trojanowski, J.Q., Rumsey, J.M., Hicks, R., Cameron, J., Chen, D., and Chen, W.G. 2011. Harnessing neuroplasticity for clinical applications. *Brain*, 134: 1591–1609.

Crick, F. 1995. *The Astonishing Hypothesis: The Scientific Search for the Soul*. New York, NY: Touchstone.

Crick, F., and Koch, C. 1990. Toward a neurobiological theory of consciousness. *Seminars in the Neurosciences*, 2: 263–275.

Crisp, T., Porter, S., and Elschof, G.T. (eds). 2016. *Neuroscience and the Soul: The Human Person in Philosophy, Science, and Theology*. Grand Rapids, MI: Eerdmans.

Davidson, D. 1963. Actions, reasons, and causes. *The Journal of Philosophy*, 60(23): 685–700.

Davidson, D. 2001. "Mental events." In Davidson, D. (ed.), *Essays on Actions and Events*, 207–224. New York, NY: Oxford University Press.

Davidson, D. 2001b. "Three varieties of knowledge." In Davidson, D. (ed.), *Subjective, Intersubjective, Objective*, 205–220. New York, NY: Oxford University Press.

De Haan, D.D. 2018. "Hylomorphism and the new mechanist philosophy in biology, neuroscience, and psychology." In Simpson, W.M.R., Koons, R.C., Nocholas, J.T. (eds.), *Neo-Aristotelian Perspectives on Contemporary Science*, 293–326. New York, NY: Routledge.

De Haan, D.D., and Meadows, G.A. 2014. Aristotle and the philosophical foundations of neuroscience. *Proceedings of the American Catholic Philosophical Association*, 87: 213–230.

Dehaene, S. 2014. *Consciousness and the Brain: Deciphering How the Brain Codes Our Thoughts*. New York, NY: Penguin.

Dehaene, S., and Changeux, J.-P. 2011. Experimental and theoretical approaches to conscious processing. *Neuron*, 70: 200–227. doi: 10.1016/j.neuron.2011.03.018

Dehaene, S., Lau, H., and Kouider, S. 2017. What is consciousness, and could machines have it? *Science*, 358: 486–492. doi: 10.1126/science.aan8871

Dennett, D. 1978. Current issues in the philosophy of mind. *American Philosophical Quarterly*, 15: 249–261.

Dennett, D.C. 1995. The unimagined preposterousness of zombies. *Journal of Consciousness Studies*, 2(4): 322–325.

des Chene, D. 2000. *Life's Form: Late Aristotelian Conceptions of the Soul*. Ithaca, NY: Cornell University Press.

Descartes, R. 1996. *Meditations on First Philosophy: With Selections from the Objections and Replies*. Edited by Cottingham, J. Cambridge, UK: Cambridge University Press.

Eberl, J.T. 2006. *Thomistic Principles and Bioethics*. New York, NY: Routledge.

Eccles, J. 1980. *The Human Psyche*. Heidelberg: Springer.

Ellis, G.F.R. 2009. "Top-down causation and the human brain." In Murphy, N., Ellis, G.F.R., and O'Connor, T. (eds.), *Downward Causation and the Neurobiology of Free Will*, 63–82. Heidelberg: Springer.

Ellis, G.F.R. 2015. "Recognizing top-down causation." In Aguirre, A., Foster, B., and Merali, Z. (eds.), *Questioning the Foundations of Physics*, 17–44. Heidelberg: Springer. doi: 10.1007/978-3-319-13045-3_3

Ellis, G.F.R. 2016. *How Can Physics Underlie the Mind: Top-Down Causation in the Human Context*. Heidelberg: Springer.

Esmaeilzadeh, M., Dictus, C., Kayvanpour, E., Sedaghat-Hamedani, F., Eichbaum, M., Hofer, S., Engelmann, G., Fonouni, H., Golriz, M., Schmidt, J., and Unterberg, A. 2010. One life ends, another begins: Management of a brain-dead pregnant mother—A systematic review. *BMC Medicine*, 8: 74. doi: 10.1186/1741-7015-8-74

Fallon, F. 2016. "Integrated information theory of consciousness." In Fieser, J., and Dowden, B. (eds.), *Internet Encyclopedia of Philosophy*. https://iep.utm.edu/int-info/

Farahany, N.A., Greely, H.T., and Giattino, C.M. 2019. Part-revived pig brains raise slew of ethical quandaries. *Nature*, 568: 299–302.

Farahany, N.A., Greely, H.T., Hyman, S., Koch, C., Grady, C., Paşca, S.P., Sestan, N., Arlotta, P., Bernat, J.L., Ting, J., and Lunshof, J.E. 2018. The ethics of experimenting with human brain tissue. *Nature*, 556: 429–432.

Feigl, H. 1967. *The Mental and the Physical: The Essay and Postscript*. Minneapolis, MN. University of Minnesota Press.

Feser, E. 2009. *Aquinas: A Beginner's Guide*. London, UK: Oneworld.

Fine, K. 2012. "Guide to ground." In Correia, F., and Schnieder, B. (eds.), *Metaphysical Grounding*, 37–80. Cambridge, UK: Cambridge University Press.

Finger, S. 2000. *Minds Behind the Brain: A History of the Pioneers and Their Discoveries*. New York, NY: Oxford University Press.

Fins, J.J. 2015. *Rights Come to Mind: Brain Injury, Ethics, and the Struggle for Consciousness*. Cambridge, UK: Cambridge University Press.

Fins, J.J., Schiff, N.D., and Foley, K.M. 2007. Late recovery from the minimally conscious state: Ethical and policy implications. *Neurology*, 68: 304–307.

Fodor, J. 1989. Making mind matter more. *Philosophical Topics*, XVII (1): 59–79.

Foster, J. 1991. *The Immaterial Self: A Defense of the Cartesian Dualist Conception of the Mind*. New York, NY: Routledge.

Foster, J.A. 1968. Psychophysical causal relations. *American Philosophical Quarterly*, 5(1): 64–70.

Frässle, S., Sommer, J., Jansen, A., Naber, M., and Einhäuser, W. 2014. Binocular rivalry: Frontal activity relates to introspection and action but not to perception. *The Journal of Neuroscience*, 34(5): 1738–1747.

Freddoso, A.J. 2012. Oh My Soul, There's Animals and Animals: Some Thomistic Reflections on Contemporary Philosophy of Mind. Paper presented at the second Thomistic workshop, Mount Saint Mary College, Newburgh, June 2012.

Frith, C.D., and Rees, G. 2017. "A brief history of the scientific approach to the study of consciousness." In Schneider, S., and Velmans, M. (eds.), *The Blackwell Companion to Consciousness*, 3–16. Oxford, UK: Wiley Blackwell.

Fumerton, R. 2013. *Knowledge, Thought, and the Case for Dualism*. Cambridge, UK: Cambridge University Press.

Garcia, R.K. 2014. Closing in on causal closure. *Journal of Consciousness Studies*, 21(1–2): 96–109.

Gertler, B. 2008. "In defense of mind-body dualism." In Feinberg, J., and Shafer-Landau, R. (eds.), *Reason and Responsibility: Readings in Some Basic Problems of Philosophy*, 285–296. Belmont, CA: Thomson-Wadsworth.

Giacino, J.T. 2004. The vegetative and minimally conscious states: Consensus-based criteria for establishing diagnosis and prognosis. *NeuroRehabilitation*, 19: 293–298.

Giacino, J.T., Fins, J.J., Laureys, S., and Schiff, N.D. 2014. Disorders of consciousness after aquired brain injury: The state of the science. *Nature Reviews Neurology*, 10: 99–114. doi: 10.1038/nrneurol.2013.279

Giacino, J.T., and Kalmar, K. 1997. The vegetative and minimally conscious states: A comparison of clinical features and functional outcome. *The Journal of Head Trauma Rehabilitation*, 12(4): 36–51.

Giacino, J.T., Kalmar, K., and Whyte, J. 2004. The JFK coma recovery scale–Revised: Measurement characteristics and diagnostic utility. *Archives of Physical Medicine and Rehabilitation*, 85(12): 2020–2029.

Gibb, S.C. 2013. "Introduction." In Gibb, S.C., Lowe, E.J., and Ingthorsson, R.D. (eds.), *Mental Causation and Ontology*, 1–16. Oxford, UK: Oxford University Press.

Gillett, C., and Loewer, B. (eds). 2001. *Physicalism and Its Discontents*. Cambridge, UK: Cambridge University Press.

Gilson, É. 2009. *From Aristotle to Darwin and Back Again: A Journey in Final Causality, Species, and Evolution*. San Francisco, CA: Ignatius Press.

Glickstein, M. 2014. *Neuroscience: A Historical Introduction.* Cambridge, MA: MIT Press.

Glüer, K. 2011. *Donald Davidson: A Short Introduction.* New York, NY: Oxford University Press.

Göcke, B.P. (ed). 2012. *After Physicalism.* Notre Dame, IN: University of Notre Dame Press.

Goetz, S. 2011. *The Causal Closure Argument.* Metanexus. http://www.metanexus.net/essay/causal-closure-argument [Accessed December 4, 2014]

Goff, P. 2019. *Galileo's Error: Foundations for a New Science of Consciousness.* New York, NY: Pantheon Books.

Goff, P., Seager, W., and Allen-Hermanson, S. 2017. "Panpsychism." In Zalta, E.N. (ed.), *The Stanford Encyclopedia of Philosophy.* Winter 2017 Edition. plato.stanford.edu/entries/panpsychism/

Goldstein, S. 2017. "Bohmian mechanics." In Zalta, E.N. (ed.), *The Stanford Encyclopedia of Philosophy.* Summer 2017 Edition. plato.standford.edu/entires/qm-bohm/

Goodrich, G.L., Martinsen, G.L., Flyg, H.M., Kirby, J., Asch, S.M., Brahm, K.D., Brand, J.M., Cajamarca, D., Cantrell, J.L., Chong, T., and Dziadul, J.A. 2013. Development of a mild traumatic brain injury-specific vision sceening protocal: A Delphi Study. *Journal of Rehabilitation Research & Development*, 50(6): 757–769.

Gosseries, O., Haibo, D., Laureys, S., and Boly, M. 2014. Measuring consciousness in severely damaged brains. *Annual Review of Neuroscience*, 37: 457–478. doi: 10.1146/annurev-neuro-062012-170339

Guta, M.P. 2019a. "Introduction." In Guta, M.P. (ed.), *Consciousness and the Ontology of Properties*, 1–12. New York, NY: Routledge.

Guta, M.P. 2019b. "The non-causal account of the spontaneous emergence of phenomenal consciousness." In Guta, M.P. (ed.), *Consciousness and the Ontology of Properties*, 126–151. New York, NY, Routledge.

Hacker, P.M.S. 2007. *Human Nature: The Categorial Framework.* Oxford, UK: Blackwell.

Haldane, J. 1998. A return to form in the philosophy of mind. *Ratio*, 11(3): 253–277.

Hart, W.D. 1988. *The Engines of the Soul.* Cambridge, UK: Cambridge University Press.

Hart, W.D. 1994. "Dualism." In Guttenplan, S. (ed.), *A Companion to the Philosophy of Mind*, 265–269. Oxford, UK: Blackwell Publishers.

Hasker, W. 1999. *The Emergent Self.* Ithaca, NY: Cornell University Press.

Hasker, W. 2010. "Persons and the unity of consciousness." In Koons, R., and Beeler, G. (eds.), *The Waning of Materialism*, 175–190. New York, NY: Oxford University Press.

Heeger, D.J., Huk, A.C., Geisler, W.S., and Albrecht, D.G. 2000. Spikes versus BOLD: What does neuroimaging tells us about neuronal activity? *Nature Neuroscience*, 3(7): 631–633.

Hill, C.S. 1991. *Sensations: A Defense of Type Materialism.* New York, NY: Cambridge University Press.

Hippocrates. 1886. *The Genuine Works of Hippocrates*. Translated by Adams, F. New York, NY: W. Wood and Company.
Hodgson, D. 1991. *The Mind Matters: Conciousness and Choice in a Quantum World*. Oxford, UK: Oxford University Press.
Hohwy, J. 2007. The search for neural correlates of consciousness. *Philosophy Compass*, 2(3): 461–474.
Hohwy, J., and Bayne, T. 2015. "The neural correlates of consciousness: Causes, confounds and constituents." In Miller, S.M. (ed.), *The Constitution of Phenomenal Consciousness: Toward a science and theory*, 155–176. Philadelphia, PA: John Benjamins Publishing Company.
Horton, M. 2011. *The Christian Faith: A Systematic Theology for Pilgrims On the Way*. Grand Rapids, MI: Zondervan.
Hume, D. 2007. *An Enquiry Concerning Human Understanding*. New York, NY: Oxford University Press.
Hunter, M.R., Gillespie, B.W., and Chen, S.Y.-P. 2019. Urban nature experiences reduce stress in the context of daily life based on salivary biomarkers. *Frontiers in Psychology*, 10: 722. doi: 10.3389/fpsyg.2019.00722
Inman, R.D. 2018. *Substances and the Fundamentality of the Familiar: A Neo-Aristotelian Mereology*. New York, NY: Routledge.
Jackson, F. 1982. Epiphenomenal qualia. *The Philosophical Quarterly*, 32(127): 127–136.
Jackson, F. 1986. What Mary didn't know. *Journal of Philosophy*, 83(5): 291–295.
Jaworski, W. 2016. *Structure and the Metaphysics of Mind: How Hylomorphism Solves the Mind-Body Problem*. New York, NY: Oxford University Press.
Jensen, M., and Overgaard, M. 2011. Neural plasticity and consciousness. *Frontiers in Psychology*, 2: 1–2. doi: 10.3389/fpsyg.2011.00191
Kandel, E.R., and Hudspeth, A.J. 2013. "The brain and behavior." In Kandel, E.R., Schwartz, J.H., Jessell, T.M., Siegelbaum, S.A., and Hudspeth, A.J. (eds.), *Principles of Neural Science*, 5–20. New York, NY: McGraw-Hill.
Keener, C.S. 2011. *Miracles: The Credibility of the New Testament Accounts*. Grand Rapids, MI: Baker Academic.
Kenny, A. 1968. *Descartes*. New York: Random House.
Kim, J. 1973. Causation, nomic subsumption, and the concept of event. *The Journal of Philosophy*, 70(8): 217–236.
Kim, J. 1993. "The non-reductivist's troubles with mental causation." In Heil, J., and Mele, A. (eds.), *Mental Causation*, 189–210. New York, NY: Oxford University Press.
Kim, J. 2000. *Mind in a Physical World*. Cambridge, MA: MIT Press.
Kim, J. 2001a. "Lonely souls: Causality and substance dualism." In Corcoran, K. (ed.), *Soul, Body, and Survival: Essays on the Metaphysics of Human Persons*, 30–43. Ithaca, NY: Cornell University Press.
Kim, J. 2001b. "Mental causation and consciousness: The two mind-body problems for the physicalist." In Gillett, C., and Loewer, B. (eds.), *Physicalism and Its Discontents*, 271–283. Cambridge, UK: Cambridge University Press.

Kim, J. 2003. "Lonely souls: Causality and substance dualism." In O'Connor, T. and Robb, D. (eds.), *Philosophy of Mind: Contemporary Readings*, 65–78. New York, NY: Routledge.

Kim, J. 2005. *Physicalism, Or Something Near Enough.* Princeton, NJ: Princeton University Press.

Kim, J. 2009. "Mental causation." In McLaughlin, B.P., Beckermann, A., and Walter, S. (eds.), *The Oxford Handbook of Philosophy of Mind*, 29–49. Oxford, UK: Oxford University Press.

Kim, J. 2011. *Philosophy of Mind.* Third Edition. Boulder, CO: Westview Press.

King, P. 2012. "Body and soul." In Marenbon, J. (ed.), *The Oxford Handbook of Medieval Philosophy*, 505–524. New York, NY: Oxford University Press.

Koch, C. 2004. *The Quest for Consciousness: A Neurobiological Approach.* Englewood, CO: Roberts and Company Publishers.

Koch, C. 2012. *Consciousness: Confessions of a Romantic Reductionist.* Cambridge, MA: The MIT Press.

Koch, C. 2019a. Is Death reversible? An experiment that partially revived slaughterhouse pig brains raises questions about the precise end point of life. *Scientific America*, October.

Koch, C. 2019b. *The Feeling of Life Itself: Why Consciousness Is Widespread but Can't Be Computed.* Cambridge, MA: MIT Press.

Koch, C., Massimini, M., Boly, M., and Tononi, G. 2016. Neural correlates of consciousness: progress and problems. *Nature Reviews Neuroscience*, 17: 307–321.

Koons, R.C., and Bealer, G. (eds). 2010. *The Waning of Materialism.* New York, NY: Oxford University Press.

Kripke, S. 1981. *Naming and Necessity.* Malden, MA: Blackwell.

Lamme, V.A.F. 2006. Towards a true neural stance on consciousness. *Trends in Cognitive Sciences*, 10(11): 494–501. doi: 10.1016/j.tics.2006.09.001

Lamme, V.A.F. 2020. Visual functions generating conscious seeing. *Frontiers in Psychology*, 11(83): 1–10. doi: 10.3389/fpsyg.2020.00083

LaRock, E. 2012. An empirical case against central state materialism. *Philosophia Christi*, 12(2): 409–426.

LaRock, E. 2013. Aristotle and agent-directed neuroplasticity. *International Philosophical Quarterly*, 53(4): 385–408.

LaRock, E., Schwartz, J., Ivanov, I., and Carreon, D. 2020. A strong emergence hypothesis of conscious integration and neural rewiring. *International Philosophical Quarterly*, 60(1): 97–115. doi: 10.5840/ipq202016146

Lau, H., and Rosenthal, D. 2011. Empirical support for higher-order theories of conscious awareness. *Trends in Cognitive Sciences*, 15(8): 365–373. doi: 10.1016/j.tics.2011.05.009

Laudan, L. 1977. *Progress and Its Problems: Towards a Theory of Scientific Growth.* Berkeley, CA: University of California Press.

Laureys, S. 2005. Death, unconsciousness and the brain. *Nature Reviews Neuroscience*, 6: 899–909.

Laureys, S., Celesia, G. G., Cohadon, F., Lavrijsen, J., León-Carrión, J., Sannita, W. G., Sazbon, L., Schmutzhard, E., von Wild, K.R., Zeman, A., and Dolce, G. 2010.

Unresponsive wakefulness syndrome: a new name for the vegetative state or apallic syndrome. *BMC Medicine,* 8: 68. doi: 10.1186/1741-7015-8-68

Lavazza, A. 2021. Potential ethical problems with human cerebral organoids: Consciousness and moral status of future brains in a dish. *Brain Research,* 1750(2021): 147146. doi: 10.1016/j.brainres.2020.147146

Lavazza, A., and Robinson, H. (eds). 2014. *Contemporary Dualism: A Defense.* New York, NY: Routledge.

Leftow, B. 2010. "Soul, mind, and brain." In Koons, R.C., and Bealer, G. (eds.), *The Waning of Materialism,* 395–415. New York, NY: Oxford University Press.

Leopold, D.A., and Logothetis, N.K. 1996. Activity changes in early visual cortex reflect monkeys' percepts during binocular rivalry. *Nature,* 379(6565): 549–553.

Licona, M.R. 2010. *The Resurrection of Jesus: A New Historiographical Approach.* Downer Groves, IL: InterVarsity Press.

Loar, B. 1990. Phenomenal States. *Philosophical Perspectives,* 4: 81–108.

Loftin, R.K., and Farris, J.R. (eds). 2018. *Christian Physicalism? Philosophical Theological Criticisms.* Lanham, MD: Lexington Books/Rowman & Littlefield.

Logothetis, N.K., and Schall, J.D. 1989. Neuronal correlates of subjective visual perception. *Science,* 245(4919): 761–763.

Loke, A. 2020. *Investigating the Resurrection of Jesus Christ: A New Transdisciplinary Approach.* New York, NY: Routledge.

Loose, J.J., Menuge, A.J.L., and Moreland, J.P. (eds). 2018. *The Blackwell Companion to Substance Dualism.* Oxford, UK: Wiley Blackwell.

Loux, M.J. 2006. *Metaphysics: A Contemporary Introduction.* Third Edition. New York, NY: Routledge.

Lowe, E.J. 1996. *Subjects of Experience.* New York, NY: Cambridge University Press.

Lowe, E.J. 1998. *The Possibility of Metaphysics: Substance, Identity, and Time.* Oxford, UK: Oxford University Press.

Lowe, E.J. 2000. Causal closure principles and emergentism. *Philosophy,* 75(4): 571–585.

Lowe, E.J. 2001. "Identity, composition, and the simplicity of the self." In Corcoran, K. (ed.), *Soul, Body, and Survival: Essays on the Metaphysics of Human Persons,* 139–158. Ithaca, NY: Cornell University Press.

Lowe, E.J. 2002. *A Survey of Metaphysics.* New York, NY: Oxford University Press.

Lowe, E.J. 2003. "Physical causal closure and the invisibility of mental causation." In Walter, S., and Heckmann, H.-D. (eds.), *Physicalism and Mental Causation,* 137–154. Charlottesville, VA: Imprint Academic.

Lowe, E.J. 2006. Non-cartesian substance dualism and the problem of mental causation. *Erkenntnis,* 65(1): 5–23.

Lowe, E.J. 2009. "Dualism." In McLaughlin, B.P., Beckermann, A., and Walter, S. (eds.), *The Oxford Handbook of Philosophy of Mind,* 66–84. New York, NY: Oxford University Press.

Lowe, E.J. 2010. "Substance dualism: A non-cartesian approach." In Koons, R.C., and Bealer, G. (eds.), *The Waning of Materialism,* 439–462. New York, NY: Oxford University Press.

Lowe, E.J. 2012. "A neo-Aristotelian substance ontology: Neither relational nor constituent." In Tahko, T.E. (ed.), *Contemporary Aristotelian Metaphysics*, 229–248. Cambridge, UK: Cambridge University Press.

Lowe, E.J. 2013. "Substance causation, powers, and human agency." In Gibb, S.C., Lowe, E.J., and Ingthorsson, R.D. (eds.), *Mental Causation and Ontology*, 153–172. Oxford, UK: Oxford University Press.

Madden, J.D. 2013. *Mind, Matter, & Nature: A Thomistic Proposal for the Philosophy of Mind*. The Catholic University of American Press.

Manning, J. 2012. Withdrawal of life-sustaining treatment from a patient in a minimally conscious state. *Journal of Law and Medicine*, 19(3): 430–435.

Maquet, P., Degueldre, C., Delfiore, G., Aerts, J., Péters, J.M., Luxen, A., and Franck, G. 1997. Functional neuroanatomy of human slow wave sleep. *Journal of Neuroscience*, 17(8): 2807–2812. doi: 10.1523/JNEUROSCIE.17-08-02807.1997

Marcus, E. 2004. Why zombies are inconceivable. *Australasian Journal of Philosophy*, 82(3): 477–490.

Marmodoro, A. 2010. "Do powers need powers to make them powerful? From pandispositionalism to Aristotle." In Marmodoro, A. (ed.), *The Metaphysics of Powers: Their Grounding and their Manifestations*, 27–40. New York, NY: Routledge.

Marmodoro, A. 2013. Aristotle's hylomorphism without reconditioning. *Philosophical Inquiry*, 36(1–2): 5–22.

Marmodoro, A. 2014a. *Aristotle On Perceiving Objects*. New York, NY: Oxford University Press.

Marmodoro, A. 2014b. "Causation without glue: Aristotle on causal powers." In Viano, C., Natali, C., and Zingano, M. (eds.), *Les Quatre Causes d'Aristotle: Origins et Interprétations*, 221–246. Leuven: Peeters.

Marmodoro, A. 2018. "Potentiality in Aristotle's metaphysics." In Engelhard, K., and Quante, M. (eds.), *Handbook of Potentiality*, 15–43. Heidelberg: Springer.

Marmodoro, A., and Page, B. 2016. Aquinas on forms, substances and artifacts. *Vivarium*, 54: 1–21.

Martini, F.H., Timmons, M.J., and Tallitsch, R.B. 2006. *Human Anatomy*. Fifth Edition. San Francisco, CA: Pearson.

Massimini, M., Ferrarelli, F., Huber, R., Esser, S.K., Harpreet, S., and Tononi, G. 2005. Breakdown of cortical effective connectivity during sleep. *Science*, 309(5744): 2228–2232. doi: 10.1126/science.1117256

Massimini, M., and Tononi, G. 2018. *Sizing Up Consciousness: Towards an Objective Measure of the Capacity for Experience*. New York, NY: Oxford University Press.

Maung, H.H. 2018. Dualism and its place in a philosophical structure for psychiatry. *Medicine, Health Care and Philosophy*, 22(1): 59–69.

Mayr, E. 2011. *Understanding Human Agency*. New York, NY: Oxford University Press.

Metzinger, T. 2000. "Introduction." In Metzinger, T. (ed.), *Neural Correlates of Consciousness: Empirical and Conceptual Questions*, 1–12. Cambridge, MA: The MIT Press.

Metzinger, T. 2003. *Being No One: The Self-Model Theory of Subjectivity*. Cambridge, MA: MIT Press.

Mogensen, J. 2011. Reorganization of the injured brain: Implications for studies of the neural substrate of cognition. *Frontiers in Psychology*, 2: 1–10.

Monti, M.M., Vanhaudenhuyse, A., Coleman, M.R., Boly, M., Pickard, J.D., Tshibanda, L., Owen, A.M., and Laureys, S. 2010. Willful modulation of brain activity in disorders of consciousness. *The New England Journal of Medicine*, 362: 579–589.

Mørch, H.H. 2017. Is matter conscious? Why the central problem in neuroscience is mirrored in physics. *Nautilus*, 47. http://nautil.us/issue/47/consciousness/is-matter-conscious

Mørch, H.H. 2019a. Is consciousness intrinsic? A problem for the integrated information theory. *Journal of Consciousness Studies*, 26(1–2): 133–162.

Mørch, H.H. 2019b. "Phenomenal knowledge why: The explanatory knowledge argument against physicalism." In Coleman, S. (ed.), *The Knowledge Argument*, 223–253. New York, NY: Cambridge University Press.

Moreland, J.P. 2005. If you can't reduce, you must eliminate. *Philosophia Christi*, 7(2): 463–473.

Moreland, J.P. 2008. *Consciousness and the Existence of God: A Theistic Argument*. New York, NY: Routledge.

Moreland, J.P. 2011. Substance dualism and the argument from self-awareness. *Philosophia Christi*, 13(1): 21–34.

Moreland, J.P. 2015. Tweaking Dallas Willard's ontology of the human person. *Journal of Spiritual Formation & Soul Care*, 8(2): 187–202.

Moreland, J.P. 2016. A critique of and alternative to Nancey Murphy's Christian physicalism. *European Journal for Philosophy of Religion*, 8(2): 107–128.

Moreland, J.P. 2018. "In defense of a thomistic-like dualism." In Loose, J.J., Menuge, A.J.L., and Moreland, J.P. (eds.), *The Blackwell Companion to Substance Dualism*, 102–122. Oxford, UK: Wiley Blackwell.

Moreland, J.P., and Craig, W.L. 2003. *Philosophical Foundations for a Christian Worldview*. Downers Grove, IL: IVP Academic.

Moreland, J.P., Meister, C., and Sweis, K.A. 2013. *Debating Christian Theism*. New York, NY: Oxford University Press.

Moreland, J.P., and Wallace, S. 1995. Aquinas versus Locke and Descartes on the human person and end-of-life ethics. *International Philosophical Quarterly*, 35(3): 319–330.

Mormann, F., and Koch, C. 2007. Neural correlates of consciousness. *Scholarpedia: the Peer-Reviewed Open-Access Encyclopedia*, 2(12): 1740. doi: 10.4249/scholarpedia.1740

Munoz-Cespedes, J.M., Rios-Lago, M., Paul, N., and Maestu, F. 2005. Functional neuroimaging studies of cognitive recovery after acquired brain damage in adults. *Neuropsychology Review*, 15(4): 169–183.

Murphy, N. 1998. "Human nature: Historical, scientific, and religious issues." In Brown, W.S., Murphy, N., and Malony, H.N. (eds.), *Whatever Happened to the*

*Soul? Scientific and Theological Portraits of Human Nature*, 1–24. Minneapolis, MN: Fortress Press.

Naci, L., Cusack, R., Anello, M., and Owen, A.M. 2014. A common neural code for similar conscious experiences in different individuals. *Proceedings of the National Academy of Sciences of the USA*, 111(39): 14277–14282.

Nagel, T. 2012. *Mind & Cosmos: Why the Materialist Neo-Darwinian Conception of Nature Is Almost Certainly False*. New York, NY: Oxford University Press.

Nair-Collins, M., and Miller, F.G. 2017. Do the "brain dead" merely appear to be alive? *Journal of Medical Ethics*, 43: 747–753.

Noë, A., and Thompson, E. 2004. Are there neural correlates of consciousness? *Journal of Consciousness Studies*, 11(1).

Noordhof, P. 1999. The overdetermination argument versus the cause-and-essence principle—No contest. *Mind*, 108(430): 367–375.

Norkowski, J.M. 2018. Brain based criteria for death in the light of the Aristotelian-Scholastic anthropology. *Scientia et Fides*, 6(1): 153–188.

O'Conner, T. 1994. Emergent properties. *American Philosophical Quarterly*, 31: 91–104.

O'Conner, T., and Robb, D. (eds.) 2003. *Philosophy of Mind: Contemporary Readings*. New York, NY: Routledge.

Oderberg, D.S. 1999. "Introduction." In Oderberg, D.S. (ed.), *Form and Matter: Themes in Contemporary Metaphysics*, vii–xi. Malden, MA: Blackwell.

Oderberg, D.S. 2005. "Hylemorphic dualism." In Paul, E.F., Miller Jr, F.D., and Paul, J. (eds.), *Personal Identity*, 70–99. New York, NY: Cambridge University Press.

Olson, E.T. 2007. *What Are We? A Study in Personal Ontology*. New York, NY: Oxford University Press.

Overgaard, M. 2017. The status and future of consciousness research. *Frontiers in Psychology*, 8: 1–4.

Overgaard, M., and Mogensen, J. 2011. A framework for the study of multiple realizations: The importance of levels of analysis. *Fontiers in Psychology*, 2: 1–10.

Owen, A.M. 2020. Improving diagnosis and prognosis in disorders of consciousness. *Brain: A Journal of Neurology*, 2020(143): 1050–1053.

Owen, A.M., Coleman, M.R., Boly, M., Davis, M.H., Laureys, S., and Pickard, J.D. 2006. Detecting awareness in the vegetative state. *Science*, 313(5792): 1402.

Owen, M. 2015. Physicalism's epistemological incompatibility with a priori knowledge. *Teorema*, XXXIV(3): 123–139.

Owen, M. 2019a. Circumnavigating the causal pairing problem with hylomorphism and the integrated information theory of consciousness. *Synthese*. doi: 10.1007/s11229-019-02403-6

Owen, M. 2019b. Exploring common ground between integrated information theory and Aristotelian metaphysics. *Journal of Consciousness Studies*, 26(1–2): 163–187.

Owen, M. 2020a. Aristotelian causation and neural correlates of consciousness. *Topoi*, 39(5): 1113–1124. doi: 10.1007/s11245-018-9606-9

Owen, M. 2020b. Cerebral organoids: Conscious subjects or zombies? *Blog of the American Philosophical Association*. https://blog.apaonline.org/2020/10/01/cerebral-organoids-conscious-subjects-or-zombies/

Owen, M. 2020c. Conscious matter and matters of conscience: An opinionated précis of The Feeling of Life Itself. *Philosophia Christi*, 22(1): 143–154.

Owen, M. 2020d. The causal efficacy of consciousness. *Entropy*, 22(8). doi: 10.3390/e22080823

Owen, M., and Guta, M.P. 2019. Physically sufficient neural mechanisms of consciousness. *Frontiers in Systems Neuroscience*, 13(24). doi: 10.3389/fnsys.2019.00024

Pan, J., Xie, Q., Qin, P., Chen, Y., He, Y., Huang, H., Wang, F., Ni, X., Cichocki, A., Yu, R., and Li, Y. 2020. Prognosis for patients with cognitive motor dissociation identified by brain-computer interface. *Brain: A Journal of Neurology*, 2020(143): 1177–1189. doi: 10.1093/brain/awaa026

Papineau, D. 1998. Mind the gap. *Noûs*, 32: 373–388.

Papineau, D. 2009. "The causal closure of the physical and naturalism." In McLaughlin, B.P., Beckermann, A., and Walter, S. (eds.), *The Oxford Handbook of Philosophy of Mind*, 53–65. Oxford, UK: Oxford University Press.

Pasnau, R. 2002a. "Commentary." In Aquinas, *The Treatise on Human Nature: Summa Theologiae 1a 75-89*, 220–378. Translated and edited by Pasnau, R. Indianapolis, IN: Hackett.

Pasnau, R. 2002b. "Intoduction." In Aquinas, *The Treatise on Human Nature: Summa Theologiae 1a 75-89*, xii–xxi. Translated and edited by Pasnau, R. Indianapolis, IN: Hackett.

Pasnau, R. 2002c. *Thomas Aquinas on Human Nature: A Philosophical Study of Summa theologiae 1a 75-89*. Cambridge, UK: Cambridge University Press.

Pasnau, R. 2012. "Philosophy of mind and human nature." In Davies, B., and Stump, E. (eds.), *The Oxford Handbook of Aquinas*, 348–368. New York, NY: Oxford University Press.

Patterson, D. 2003. Review of Bennett, M.R., and Hacker, P.M.S. Philosophical foundations of neuroscience. *Notre Dame Philosophical Reviews*. https://ndpr.nd.edu/news/philosophical-foundations-of-neuroscience/

Penfield, W. 1975. *The Mystery of the Mind: A Critical Study of Consciousness and the Human Brain.* Princeton, NJ: Princeton University Press.

Plantinga, A. 1984. Advice to christian philosophers. *Faith and Philosophy*, 1(3): 253–271.

Plantinga, A. 2000. *Warranted Christian Belief.* New York, NY: Oxford University Press.

Plantinga, A. 2007. "Materialism and christian belief." In van Inwagen, P., and Zimmerman, D. (eds.), *Persons: Human and Divine*, 99–141. New York, NY: Oxford University Press.

Plantinga, A. 2008. What is "intervention"? *Theology and Science*, 6(4): 369–401.

Plantinga, A. 2012. A new argument against materialism. *Philosophia Christi*, 14(1): 9–27.

Polgar, T.W. 2011. Are sensations still brain processes. *Philosophical Psychology*, 24(1): 1–21.

Polonsky, A., Blake, R., Braun, J., and Heeger, D.J. 2000. Neuronal activity in human primary visual cortex correlates with perception during binocular rivalry. *Nature Neuroscience*, 3(11): 1153–1159.

Popper, K., and Eccles, J.C. 1983. *The Self and Its Brain: An Argument for Interactionism.* New York, NY: Routledge.

Pruss, A. 2010. Causal closure of the physical. *Alexander Pruss's Blog.* http://alexanderpruss.blogspot.com/2010/08/causal-closure-of-physical.html

Pruss, A. 2013a. Aristotelian forms and laws of nature. *Analiza i Egzystencja*, 24: 115–132.

Putnam, H. 1967. "Psychological predicates." In Capitan, W.H., and Merrill, D.D. (eds.), *Art, Mind, and Religion*, 37–48. University of Pittsburgh Press.

Rea, M.C. 2011. Hylomorphism reconditioned. *Philosophical Perspectives*, 25(1): 341–358.

Reardon, S. 2019. Pig brains kept alive for hours outside body. *Nature*, 568: 283–284. doi: 10.1038/d41586-019-01216-4

Reardon, S. 2020. Can lab-grown brains become conscious? *Nature*, 586: 658–661. doi: 10.1038/d41586-020-02986-y

Rickabaugh, B. 2018. The primacy of the mental: From Russellian Monism to substance dualism. *Philosophia Christi*, 20(1): 31–41.

Robinson, H. 1982. *Matter and Sense: A Critique of Contemporary Materialism.* New York, NY: Cambridge University Press.

Robinson, H. (ed.). 1993. *Objections to Physicalism.* New York, NY: Oxford University Press.

Robinson, H. 2012. "Qualia, qualities, and our conception of the physical world." In Göcke, B.P. (ed.), *After Physicalism*, 231–263. Notre Dame, IN: University of Notre Dame Press.

Robinson, H. 2016. *From the Knowledge Argument to Mental Substance: Resurrecting the Mind.* New York, NY: Cambridge University Press.

Robinson, H. 2020. "Dualism." In Zalta, E.N. (ed.), *The Stanford Encyclopedia of Philosophy.* Fall 2020 Edition. https://plato.stanford.edu/entries/dualism/

Ryle, G. 1949. *The Concept of the Mind.* 60th Anniversary Edition. New York, NY: Routledge.

Sandrone, S., Bacigaluppi, M., Galloni, M.R., Cappa, S.F., Moro, A., Catani, M., Filippi, M., Monti, M.M., Perani, D., and Martino, G. 2014. Weighing brain activity with the balance: Angelo Mosso's original manuscripts come to light. *Brain: A Journal of Neurology*, 137: 621–633.

Schnakers, C., Edlow, B.L., Chatelle, C., and Giacino, J.T. 2016. "Minimally conscious state." In Laureys, S., Gosseries, O., and Tononi, G. (eds.), *The Neurology of Consciousness: Cognitive Neuroscience and Neuropathology*, 167–185. Second Edition. San Deigo, CA: Academic Press.

Schnakers, C., Laureys, S., and Boly, M. 2013. "Neuroimaging of consciousness in the vegetative and minimally conscious states." In Cavanna, A.E., Nani, A., Blumenfeld, H., and Laureys, S. (eds.), *Neuroimaging of Consciousness*, 117–131. Heidelberg: Springer.

Schnakers, C., Vanhaudenhuyse, A., Giacino, J., Ventura, M., Boly, M., Majerus, S., Moonen, G., and Laureys, S. 2009. Diagnostic accuracy of the vegetative and minimally conscious state: Clinical consensus versus standardized neurobehavioral assessment. *BMC Neurology*, 9(35). doi: 10.1186/1471-2377-9-35

Searle, J.R. 1992. *The Rediscovery of the Mind.* Cambridge, MA: MIT Press.
Sheinberg, D.L., and Logothetis, N.K. 1997. The role of temporal cortical areas in perceptual organization. *Proceeding of National Academy of Science of the USA*, 94: 3408–3413.
Shewmon, D.A. 1997. Recovery from "brain death": A neurologist's apologia. *The Linacre Quarterly*, 64(1): 31–96.
Shields, C. 2014. "Aristotle." In Zalta, E.N. (ed.), *Stanford Encyclopedia of Philosophy*. Spring 2014 Edition. https://plato.stanford.edu/entries/aristotle/
Shoemaker, S. 1995. "Physicalism." In Audi, R. (ed.), *The Cambridge Dictionary of Philosophy*. New York, NY: Cambridge University Press.
Shulman, R.G. 2013. *Brain Imaging: What It Can (and Cannot) Tell Us About Consciousness.* New York, NY: Oxford University Press.
Smart, J.J.C. 1959. Sensations and brain processes. *The Philosophical Review*, 68(2): 141–156.
Smart, J.J.C. 2007. "The mind/brain identity theory." In Zalta, E.N. (ed.), *The Stanford Encyclopedia of Philosophy*. Winter 2014 Edition. https://plato.stanford.edu/entries/mind-identity/
Smith, Q. 2002. "Time was created by a timeless point: An atheist explanation of spacetime." In Ganssle, G.E., and Woodruff, D.M. (eds.), *God and Time: Essays on the Divine Nature*, 95–128. New York, NY: Oxford University Press.
Smith, R.S. 2016. *Naturalism and Our Knowledge of Reality.* New York, NY: Routledge.
Smythies, J.R., and Beloff, J. (eds). 1989. *The Case for Dualism.* Charlottesville, VA: University of Virginia Press.
Storm, J.F., Boly, M., Casali, A.G., Massimini, M., Olcese, U., Pennartz, C.M., and Wilke, M. 2017. Consciousness regained: Disentangling mechanisms, brain systems, and behavioral responses. *Journal of Neuroscience*, 37(45): 10882–10893.
Stump, E. 2003. *Aquinas.* New York, NY: Routledge.
Stump, E. 2012. "God's simplicity." In Davies, B. (ed.), *The Oxford Handbook of Aquinas*, 135–146. New York, NY: Oxford University Press.
Sturgeon, S. 1998. Physicalism and overdetermination. *Mind*, 107(426): 411–432.
Swinburne, R. 1986. *The Evolution of the Soul.* Revised Edition. New York, NY: Oxford University Press.
Swinburne, R. 1996. *Is There a God?* New York, NY: Oxford University Press.
Swinburne, R. 2003. *The Resurrection of God Incarnate.* New York, NY: Oxford University Press.
Swinburne, R. 2013. *Mind, Brain, and Free Will.* Oxford, UK: Oxford University Press.
Tahko, T.E. 2012. *Contemporary Aristotelian Metaphysics.* Cambridge, UK: Cambridge University Press.
Tanabe, S., Huang, Z., Zhang, J., Chen, Y., Fogel, S., Doyon, J., Wu, J., Xu, J., Zhang, J., Qin, P., and Wu, X. 2020. Altered global brain signal during physiological, pharmacologic, and pathologic states of unconsciousness in humans and rats. *Anesthesiology*, 132(6): 1392–1406. doi: 10.1097/ALN.0000000000003197
Teasdale, G., and Jennett, B. 1974. Assessment of coma and impaired consciousness. *The Lancet*, 304(7872): 81–84. doi: 10.1016/S0140-6736(74)91639-0

Tiehen, J. 2012. Psychophysical reductionism without type identities. *American Philosophical Quarterly*, 49(3): 223–236.

Tiehen, J. 2015. Grounding causal closure. *Pacific Philosophical Quarterly*, 97(4): 501–522. doi: 10.1111/papq.12126

Tononi, G. 2004. An information integration theory of consciousness. *BMC Neuroscience*, 5(42). doi: 10.1186/1471-2202-5-42

Tononi, G. 2017. "The integrated information theory of consciousness: An outline." In Schneider, S., and Velmans, M. (eds.), *The Blackwell Companion to Consciousness*, 243–256. Oxford, UK: Wiley Blackwell.

Tononi, G., Boly, M., Massimini, M., and Koch, C. 2016. Integrated information theory: From consciousness to its physical substrate. *Nature Reviews Neuroscience*, 17: 450–461.

Tononi, G., and Koch, C. 2015. Consciousness: here, there and everywhere? *Philosophical Transactions of The Royal Society B*, 370. doi: 10.1098/rstb.2014.0167

Tsuchiya, N., and Koch, C. 2005. Continuous flash suppression reduces negative afterimages. *Nature Neuroscience*, 8: 1096–1101. doi: 10.1038/nn1500

Tsuchiya, N., Wilke, M., Frässle, S., and Lamme, V.A. 2015. No-report paradigms: Extracting the true neural correlates of consciousness. *Trends in Cognitive Sciences*, 19(12): 757–770.

Tye, M. 1986. The subjective qualities of experience. *Mind*, 95(377): 1–17.

van Inwagen, P. 1990. *Material Beings*. Ithaca, NY: Cornell University Press.

van Inwagen, P. 1993. Précis of material beings. *Philosophy and Phenomenological Research*, LIII(3): 683–686.

van Inwagen, P. 2007. "A materialist ontology of the human person." In van Inwagen, P., and Zimmerman, D. (eds.), *Persons: Human and Divine*, 199–215. New York, NY: Oxford University Press.

Verheijde, J.L., Rady, M.Y., and Potts, M. 2018. Neuroscience and brain death controversies: The elephant in the room. *Journal of Religion and Health*, 57: 1745–1763.

Vrselja, Z., Daniele, S.G., Silbereis, J., Talpo, F., Morzov, Y.M., Sousa, A.M.M., Tanaka, B.S., Skarica, M., Pletikos, M., Kaur, N., Zhuang, Z.W., Liu, Z., Alkawadri, R., Sinusas, A.J., Latham, S.R., Waxman, S.G., and Sestan, N. 2019. Restoration of brain circulation and cellular functions hours post-mortem. *Nature*, 568: 336–343. doi: 10.1038/s41586-019-1099-1

Wachter, D.v. 2006. "Why the argument from causal closure against the existence of immaterial things is bad." In Koskinen, H.J., Vilkko, R., and Philström (eds.), *Science: A Challenge to Philosophy?*, 113–124. Frankfurt: Peter Lang.

Walls, J., and Dougherty, T. 2018. *Two Dozen (or so) Arguments for God*. New York, NY: Oxford University Press.

Weiler, N. 2020. Not "brains in a dish": Cerebral organoids flunk comparison to developing nervous system. University of California San Francisco. https://www.ucsf.edu/news/2020/01/416526/not-brains-dish-cerebral-organoids-flunk-comparison-developing-nervous-system [Accessed October 28, 2020]

Wheatstone, C. 1838. Contributions to the physiology of vision—Part the first. On some remarkable, and Hitherto unobserved, phenomena of binocular vision.

*Philosophical Transactions of the Royal Society*, 128: 371–94. doi: 10.1098/rstl.1838.0019.

Willyard, C. 2015. Rise of the organoids: Biologists are building banks of mini-organs, and learning a lot about human development on the way. *Nature*, 523(7562): 520–523.

Won, C. 2019. Mental causation as joint causation. *Synthese*. doi: 10.1007/s11229-019-02378-4

Wright, N.T. 2003. *The Resurrection of the Son of God*. Minneapolis, MN: Fortress Press.

Wuellner, B. 1956. *Dictionary of Scholastic Philosophy*. Milwaukee, WI: The Bruce Publishing Company.

Zimmerman, D. 2011. "From experience to experiencer." In Baker, M.C., and Goetz, S. (eds.), *The Soul Hypothesis*, 168–196. New York, NY: Continuum.

# Index

Note: Page locators in italics refer to figures.

accidental forms, 92–93, 95
active powers, 143–45, 148–50, 163–64
affirming the consequent, 71n7
*After Physicalism* (Göcke), 8
aggregate: ontology of, 90; substance versus, 89–92
Allen Institute for Brain Science, 43n8
Alter, T., 8
*American Philosophical Quarterly* (journal), 9
American Psychological Association, 1
angels, God and, 183
anomalous monism, 47–48
Anselm the dog example, 20, 111n21
apple trees example, 83, 93–95
Aquinas, Thomas, 2, 17; brain and, 112n24; on brain and thought, 146–47; *Disputed Questions on Spiritual Creatures*, 104; on form, 154–55; powers of soul and, 151; *Questions on the Soul*, 110n5; on rational thought and body, 146; with soul and body, 149, 151–52, 165, 183, 193n15; on soul and human embryo, 133–34, 137n25; *Summa Theologiae*, 99, 100, 105–6, 107, 111n16, 147, 151, 166n14; Thomism, 3, 88, 103; "Treatise on Human Nature," 99, 100; "Treatise On the One God," 131. *See also* neo-Thomistic hylomorphism
*Are We Bodies or Souls?* (Swinburn), 10
Aristotelianism, 88, 95, 130, 140
Aristotelian powers ontology, 142–45
Aristotelian terminology: en-forming relation and, 89, 96–99; form, 89, 92–95; hylomorphism and, 89; material substance and, 89, 95–96; matter, 89, 95; neo-Thomistic hylomorphism and, 87, 88–99; substance versus aggregate, 89–92
Aristotle: with active and passive powers, 143, 145, 150; causation, 3, 5, 14, 141, 156, 183, 191; consciousness and, 2; *dunamis* and, 166n8; *Metaphysics*, 142–43; *On the Soul*, 5, 16n10, 101; *Physics*, 133, 143, 144; on rationality and humanity, 146
Arnauld, Antoine, 74
artificial stimulation, consciousness and, 187–89

Association for the Scientific Study of Consciousness, 18
assumptions, spatial relations and unstated, 121–22
*The Astonishing Hypothesis* (Crick), 17
Augustine, 101, 107, 111n15

Baars, Bernard, 16n8, 178
Bacon, Francis, 133
Bailey, Andrew, 8, 81
Baker, Lynne Rudder, 9
Baker, M. C., 8
Bannister, Roger, 45
Bayne, Tim, 9
Bealer, George, 8
Beckham, David, 97
bee sting: consciously feeling, 6; conscious sensation of, 6, 157, 158–59; MBP to sense, *160, 161*
behavioral response, with DOC, 173
beliefs: CCP and rational, 68, 70; inconsistent, 123–24; spatial relations with inconsistent, 122–25
Beloff, J., 7
Bennett, M. R., 162
Berger, Hans, 23
Bertone, 22–23
Bhaduri, Aparna, 44n24
the Bible, 100
Big Bang, 66, 81–82
binocular rivalry, 22, 24–26
biological laws, 82, 83, 160
biological regularities: examples of, 154; FBR, 155, 156, 158; MBP model of NCC and, 153–56
Blackwell, 8
*The Blackwell Companion to Substance Dualism* (Craig and Moreland), 9
Block, Ned, 29, 31–32, 43n12
bodily/body power, 193n19
bodily events, mental and, 45
body: Cartesian dualism and, 104; corpse, 188–89, 194n22; form of, 92, 106–7; as inseparable aspect of substance, 104; MBP, 148–53; mind and, 76–77, 78, 85n1, 121, 126–28, 140–41; mind-body dependence, 145–48; Philosophers Pyramid, 90; rational thought and, 146; soul and, 93, 98–99, 101–3, 106–7, 111n11, 113, 117, 126–31, 134–35, 135n4, 140–41, 147, 149, 151–52, 154, 165, 183, 193n15; as substance, 91. *See also* Mind-Body Powers, and full NCC (neural correlates of consciousness)
Bogardus, Tomas, 9
Bohm, David, 68
Boly, Mélanie, 2, 174
Bonjour, Laurence, 8, 9
bowling ball example, 143
brain, 21; Allen Institute for Brain Science, 43n8; Aquinas and, 112n24; of Bertone, 22–23; global signal, 192n6; "Measuring Consciousness in Severely Damaged Brains," 3; *Mind, Brain, & Free Will*, 9; neuronal response, 174–75; neurons, 36, 164, 165, 174, 178, 184–86, 193n20; neuroplasticity and, 34, 155, 160–61; organoids, 36, 44n25; oxygen atoms and, 60; pain and, 153; pig, 187; process theory, 30; *Sensations and Brain Processes*, 29; soul and, 41; thought and, 146–47; zombies, 36
Broad, C. D., 82
Broglie, Louis de, 68
Broglie-Bohm theory, 68
Bunge, Mario, 37

Cajal, Santiago Ramón y, 183
*The Cambridge Dictionary of Philosophy*, 4
Cambridge University Press (CUP), 7, 8
Cartesian dualism, 85, 111n18, 133; body and, 104; causation and, 28; humans and, 100; hylomorphism and, 104, 116; Kim and, 77, 111n11, 127; mind and, 126; Oderberg on, 87

*The Case for Dualism* (Smythies and Beloff), 7
causal closure, denying, 54–55
causal closure principle (CCP, causal closure of the physical domain): ambiguity of, 57–61; denial of, 52, 55, 71; rational belief and, 68, 70; substance dualism and, 46, 73; truth of, 57, 62, 66
causal exclusion problem: with causal closure, denying, 54–55; CCP and, 73; with closure, analyzing justification for, 57–66; with closure, rationally denying, 55–57; with closure as false, 66–70; mental causation and, 51–70; with overdetermination, positing, 53–54
causality, spatiality and, 82
causal pairing, en-forming: body, soul and, 113; hylomorphic responses, 115–19; with hylomorphic solution, proposed, 125–28; objections, 129–35; presupposed relation, 132–35; relation, 126–28, 133; spatial relations, questioning necessity of, 119–25
causal pairing problem, 43n14, 85, 86n10; dualism and, 42, 54, 71, 135; hylomorphism and, 10, 113, 139; Kim and, 74, 78–84, 86n10, 114–19; neo-Thomistic hylomorphism and, 14, 73, 87, 110, 113; reconsidering, 114–25; reduxed, 129–32; relation, 74–75; soul-body unity and, 102; spatiality, necessity of, 75–77; substance dualism and, 8, 11, 12, 46, 126; summarizing, 74–78
causal pairing relations: God and, 132; metaphysical explanation and, 136n16; mind-body unity and, 127, 128; spatial and, 76–77, 114, 117–18, 121–24, 126
causation: Aristotelian, 3, 5; Cartesian dualism and, 28; understanding of, 136n13. *See also* mental causation

causes, of physical events, 67–68
CCP. *See* causal closure principle
C-fiber activation: NCC and, 17–18, 42n3; pain and, 20–21, 30, 33, 42n3, 43n13
Chalmers, David, 8; lesion studies and, 42n7; MBP model and definition of, 158–62; NCC and, 18–20, 22, 28; on zombies, 35
change: active power to produce, 143–44; passive power to undergo, 143–44
check, MBP model of NCC to sign, 162, *164*
Chisholm, Roderick, 7, 39, 56
chocolate, 34, 183, 184
*The Christian Faith* (Horton), 17
*Christian Physicalism*, 10
Clemens, Samuel, 31–32
Clinton, Bill, 184
Clinton, Hillary, 103, 135
closure: analyzing justification for, 57–66; causal, 57–61; with empirical investigation, inadequacy of, 61–64; as false, reasons, 66–70; physical causal, 54, 62, 63; rationally denying, 55–57; with science, inevitable success of, 64–66
coincident objects, 85n3
*Columbia College* (McKeen), 1
coma, 3, 171
*A Companion to Philosophy of Mind* (Hart), 8
completeness-of-physics (COP), 52, 56
composition, IIT and, 178–79
conscious mind: consciousness, detecting, 3–5; with defeaters of dualism defeated, 10–14; with dualism, 6–10; interdisciplinary interest and, 5–6; physicalism and, 1, 2; presuppositions, 14–15
*The Conscious Mind* (Chalmers), 8
consciousness: Aristotle and, 2; artificial stimulation and, 187–89; Association for the Scientific Study of

Consciousness, 18; caveats, 190–91; contents of, 19, 22; defined, 170–71; detecting, 3–5; diagnosing disorders of, 171–76; experience and, 171; full NCC and, 177–82; indirectly measuring, 176–77; with MBP and full NCC, 182–87, *186*; *Neural Correlates of Consciousness*, 28; Phi and, 179, 180–82; physicalism and, 169. See also disorders of consciousness (DOC); Mind-Body Powers (MBP) model of NCC (neural correlates of consciousness); neural correlates of consciousness (NCC)

*Consciousness and the Mind-Body Problem* (Alter and Howell), 8

*Consciousness and the Ontology of Properties* (Guta), 10, 44n31

*Consciousness* (Koch), 8–9, 41

contents of consciousness, 19, 22

Continuum, 8

Cooper, John W., 42n2

COP. *See* completeness-of-physics

Copan, P., 72n14

Copenhagen interpretation, quantum mechanics of, 68

Corcilius, Klaus, 166n8

Cornell University Press, 8

corpse, 188–89, 194n22

cortisol, measuring, 177

Craig, W. L., 9, 71n11, 72n14

Crane, Tim, 9

Crick, Francis, 4, 8, 17, 24, 39

CUP. *See* Cambridge University Press

data scope, simplicity argument, 38–40

Davidson, Donald, 47, 48, 50–51

David statue example, 131

*De Anima*. *See On the Soul* (*De Anima*) (Aristotle)

Dennett, Daniel, 6, 7

Descartes, René, 82, 87, 99, 133, 188; interactionist dualism and, 6, 11, 74, 115; mind, body and, 77, 85n1, 121, 126; soul and, 41, 111n11; substance dualism and, 108, 109

Des Chene, Dennis, 141

Di, Haibo, 2

direct measurements, 176

disorders of consciousness (DOC): behavioral response, 173; coma, 3, 171; diagnosing, 171–76; MCS, 171–72, 192n1; neuronal response, 174–75; passive paradigm, 175–76; VS, 3, 171

*Disputed Questions on Spiritual Creatures* (Aquinas), 104

DOC. *See* disorders of consciousness

Dougherty, T., 72n14

DUAL, 52, 53, 61

dualism: against, 28–41; causal pairing problem and, 42, 54, 71, 135; conscious mind and, 6–10; defeating defeaters of, 10–14; identity theory's viability and, 31–37; interactionist, 6, 11, 74, 115; materialism and, 29, 42n2, 109; misrepresented, 40–41; NCC and, 12, 17–18, 41–42; neuroscience and, 17; physicalism and, 5–6, 7, 8, 18; Platonic, 100; property, 8–9, 12–13, 15, 34, 53, 104, 108–9; reconsideration of, 7; simplicity argument, 30–31; simplicity argument analysis, 37–41. *See also* Cartesian dualism; substance dualism

*Dualism in the Twenty-First Century* (conference), 10

dualistic holism, 42n2

*dunamis*, 166n8

Dylan, Bob, 4

Eberl, Jason, 190

Eccles, John, 7, 41

*Ecclesiastes*, 100

EEG. *See* electroencephalogram

efficient cause, 133, 134, 140, 153, 154, 167n19

electroencephalogram (EEG), 23, 25, 187
Elisabeth of Bohemia (Princess), 11, 41–42, 74
Ellis, George F. R., 72n13, 85n5
*The Emergent Self* (Hasker), 8
"An Empirical Case Against Central State Materialism" (LaRock), 8
empirical investigation: argument, 62–63, 64; inadequacy of, 61–64
en-forming relation: Aristotelian terminology and, 89, 96–99; body and soul, 103, 106, 117; causal pairing and, 126–28, 133; grounding and, 111n9; substantial form and, 98. *See also* causal pairing, en-forming; form
*The Engines of the Soul* (Hart), 7
en-mattered form, 130, 136n19
epiphenomenalism, 6, 11, 54, 56, 124
"Epiphenomenal Qualia" (Jackson), 34
*The Evolution of the Soul* (Swinburne), 7, 15
exclusion, IIT, 179
experience, consciousness and, 171

Farahany, Nita A., 36
FBR. *See* Forming Biological Regularities
Felix, Allyson, 148–51
Feser, Edward, 109
final cause, 133
fMRI. *See* functional magnetic resonance imaging
Fodor, Jerry, 45
form: accidental, 92–93, 95; Aquinas on, 154–55; Aristotelian terminology and, 89, 92–95; of body, 92; en-forming relation and, 89, 96–99; en-mattered, 130, 136n19; metaphysical explanation and, 110n7. *See also* neo-Thomistic hylomorphism; substantial form
formal cause, 133, 134

Forming Biological Regularities (FBR), 155, 156, 158
"Form, Structure and Hylomorphism" (Marmodoro and Paolini Poaletti), 137n26
Foster, J. A., 86n10
Foster, John, 7, 8, 71n2
Francis (Pope), 115
Frässle, Stefan, 26
*From the Knowledge Arguments to Mental Substance* (Robinson, H.), 9
Fumerton, Richard, 9
functional magnetic resonance imaging (fMRI), 23–26

Garcia, Robert, 71n4
Gassendi, Pierre, 74
*Genesis*, 100
Gertler, Brie, 8
Gibb, Sophia, 58
Gifford Lectures, 7
Gillett, C., 8
global brain signal, 192n6
Global Neuronal Workspace theory, 177–78
Göcke, B. P., 8
God, 148; angels and, 183; causal pairing relations and, 132; existence of, 69; spatiality and, 136n10
Goetz, Stewart, 8, 71n10
Gosseries, Olivia, 2
Grasso, Matteo, 167n31
grounding relation, 96, 98, 99, 106, 110, 111n9
guns firing example, 75
Guta, Mihretu, 10, 44n31, 188

Hacker, P. M. S., 162
Haldane, John, 6, 141, 154, 167n22
halfpipe example, 96
Hart, W. D., 7–8
Hasker, William, 8, 15, 16n18
Hendel, Charles, 65
Hill, Christopher, 29, 40

Hippocrates, 17
Hodgson, David, 7
Hohwy, Jakob, 28–29
Horn, L. V., 8
Horton, Michael, 17, 42n2
Howell, R. J., 8
human embryo: brain organoids from, 36; soul and, 133–34, 137n25
humanity, rationality and, 146
*The Human Psyche* (Eccles), 7
humans: Cartesian dualism and, 100; as material substance, 95–96; ontology and nature of, 99–105
"Hylemorphic Dualism" (Oderberg), 8
hylomorphism, 89, 105, 116, 125–28, 135n1, 137n26; history, recent, 140–42. *See also* accidental forms; neo-Thomistic hylomorphism; substantial forms

identity theory's viability: dualism and, 31–37; knowledge of qualia and, 34–35; multiple realizability and, 33–34; zombie nervous system and, 35–37
IIT. *See* Integrated Information Theory
immaterial cause, 120
*The Immaterial Self* (Foster, J.), 7
impact-of-the-mental (IMP), 52–54, 56, 61
impossibility: logical, 79, 80–82; metaphysical, 79, 84; nomological, 79, 82–84; physical, 79–80; relevant senses of, 79
"In Defense of Mind-Body Dualism" (Gertler), 8
indirect measurements, 176–77
Inman, Ross, 193n15
inseparable aspect, of substance, 104, 111n17
Institute of Molecular Biotechnology, Vienna, 36
Integrated Information Theory (IIT), 2, 10, 33–34, 44n19, 169–70, 177–79, 193nn11–12
integration, IIT, 179

interactionist dualism, 6, 11, 74, 115
interdependence, MBP and, 148–49
Interdependent Partner-Powers (IPP), 145
interdisciplinary interest, conscious mind and, 5–6
internal unity, 89, 92
intrinsic existence, IIT and, 178
IPP. *See* Interdependent Partner-Powers

Jackson, Frank, 34, 181
Jaworski, William, 141, 156
Jesus of Nazareth, resurrection of, 70
Jordan, Michael, 163

Kalām cosmological argument, 71n11
Karma, 81
Keener, C. S., 72n15
Kim, Jaegwon, 6, 8, 10–11, 31, 41, 43n14, 46, 165n7; with argument reconstructed and pairing problem, 78; with beliefs, inconsistent, 123–24; Cartesian dualism and, 77, 111n11, 127; causal pairing problem and, 74, 78–84, 86n10, 114–19; CCP and, 58, 62; closure and, 55; with mind-body unity, 85n1; multiple realizability and, 33; non sequitur and, 125; pain and, 136n12; pairing relation and, 74–75; physical causal closure and, 54; with premises 2 and 3, case for, 119–21; on qualia, 124; spatiality and, 75–77; thesis, analysis of, 78–84
Kim's thesis: analysis of, 78–84; logically impossible?, 80–82; metaphysically impossible?, 84; nomologically impossible?, 82–84; physically impossible?, 79–80
*Knowledge, Thought, and the Case for Dualism* (Fumerton), 9
Koch, Christof: C-fibers and, 42n3; NCC and, 4, 24, 39, 42n6, 43n8, 165n7, 167n27; physicalism and, 8–9; substance dualism and, 41

Koons, Robert C., 8, 9
Kripke, Saul, 167n28

Lancaster, Madeline, 36
LaRock, Eric, 8
Laureys, Steven, 2, 174
laws: biological, 82, 83, 160; of noncontradiction, 79; psychophysical, 47–51, 70–71; strict, 47, 48, 50
Leibniz's law, 33
lesion studies, 42n7
Lexington Books, 10
Licona, M. R., 72n15
locked-in syndrome (LIS), 173, 192n4
Loewer, B., 8
logical impossibility, 79–82
Loke, A., 72n15
Lowe, E. J., 8, 9, 15, 58, 63, 141, 165n4

macrophysical, 59–61
Madonna (singer), 4
Manchester United example, 97
Marmodoro, Anna, 136n19, 137n26, 142, 167n31, 183; on active and passive powers, 143, 145, 150; on change, 144
material cause, 133
*Material Girl*, 4
materialism: criticism of, 7; dualism and, 29, 42n2, 109; physicalism and, 1, 4, 8, 9
material substance, 89, 95–96
matter: Aristotelian terminology and, 89, 95; prime, 95. *See also* neo-Thomistic hylomorphism
*Matter and Sense* (Robinson, H.), 7
Maung, Hane Htut, 10
McKeen, J., 1
MCS. *See* minimally conscious state
measurements, 1, 3; consciousness and indirect, 176–77; of cortisol, 177; direct, 176
"Measuring Consciousness in Severely Damaged Brains" (Gosseries, Di, Laureys and Boly), 3

*Medicine, Health Care and Philosophy* (journal), 10
mental causation: causal exclusion problem and, 51–70; with psychophysical laws, lack of, 47–51; substance dualism and, 45–46
mental events: bodily and, 45; with macrophysical effects, 61; physical and, 47–49, 51–53, 55, 56, 61, 69
mental, IMP, 52–53
"Mental Measurement" (Cattell), 1
"Mental Powers" (Grasso and Marmodoro), 167n31
Menuge, Angus, 10
metaphysical explanation: causal pairing relation and, 136n16; David statue example and, 131; form and, 110n7; halfpipe example and, 96; Manchester United example and, 97; natural and, 110n7, 136n22; of NCC, 5, 142, 156, 165, 182, 184, 191
metaphysical impossibility, 79, 84
*Metaphysics* (Aristotle), 142–43
metaphysics, philosophy and, 88
methodological naturalism, 64–66
Metzinger, Thomas, 28
microphysical, physical and, 59, 61
Miller, Fred D., 8
mind: body and, 76–77, 78, 85n1, 121, 126–28, 140–41; Cartesian dualism and, 126; *A Companion to Philosophy of Mind*, 8; connotations, 13; *The Conscious Mind*, 8; *The Mystery of the Mind*, 65; *The New Directions in the Study of the Mind*, 9; *The Oxford Handbook of Philosophy of Mind*, 74; *The Rediscovery of the Mind*, 1, 4; "A Return to Form in the Philosophy of Mind," 6. *See also* conscious mind; Mind-Body Powers, and full NCC (neural correlates of consciousness)
Mind-Body Powers (MBP), and full NCC (neural correlates of consciousness), 182–87, *186*

Mind-Body Powers (MBP) model of NCC (neural correlates of consciousness): Aristotelian powers ontology and, 142–45; biological regularities and, 153–56; in context, 139–40; definition of Chalmers, 158–62; example I, 157–58; example II, 162–65; with hylomorphic history, recent, 140–42; interdependence, 148–49; mind-body dependence, 145–48; neo-Thomistic hylomorphism and, 3, 5, 153; ontological extension, 150; RPP and, 152, 156, 158; to sense bee sting, *160*; to sense bee sting, alternative system, *161*; to sign check, 162, *164*
mind-body problem, 8, 39
*Mind, Brain, & Free Will* (Swinburne), 9
*The Mind Matters* (Hodgson), 7
minimally conscious state (MCS), 171–72, 192n1
minimal neural system, 20–21, 159
monism, 9, 47–48
Mørch, Hedda Hassel, 11, 180
Moreland, J. P., 9, 10, 44n30, 72n14, 81, 136n14, 165n7, 167n22; hylomorphic history and, 140–41; organicism and, 153; on soul and body, 154
Mosso, Angelo, 22–23
motor movement, 3, 173–74
MPB. *See* Mind-Body Powers, and full NCC (neural correlates of consciousness)
multiple realizability, 6, 32–34, 44n18, 149, 160–61
Murphy, Nancy, 17
*The Mystery of the Mind* (Penfield), 65

natural explanation, metaphysical and, 110n7, 136n22
naturalism: methodological, 64–66; ontological, 64
nature, human, 99–105
*Nature* (magazine), 36

NCC. *See* Mind-Body Powers, and full NCC (neural correlates of consciousness); neural correlates of consciousness (NCC)
neo-Thomistic hylomorphism: Aristotelian terminology and, 87, 88–99; causal pairing problem and, 10, 14, 73, 87, 110, 113, 139; defined, 110; human ontology and, 99–105; incarnate souls and, 105–9; MBP model of NCC, 3, 5, 153; substance dualism and, 12
*Neural Correlates of Consciousness* (Metzinger), 28
neural correlates of consciousness (NCC): with causation and human nature, 2; C-fiber activation and, 17–18, 42n3; content-specific, 19, 42n3, 177, 185–86, 193n19; defined, 18–22, 42n6; against dualism, 28–41; dualism and, 12, 17–18, 41–42; full, 177–82; identifying, 22–26; identity theory's viability, 31–37; implications of, 26–28; Koch and, 4, 24, 39, 42n6, 43n8, 165n7, 167n27; metaphysical explanation of, 5, 142, 156, 165, 182, 184, 191; simplicity argument analysis, 37–41; variations, 43n8. *See also* Mind-Body Powers, and full NCC (neural correlates of consciousness)
neuronal response, with DOC, 174–75
neuron doctrine, 183
neurons: axon, 166n9; bee sting and, 160; brain, 36, 164, 165, 174, 178, 184–86, 193n20; firing, 149–50; Global Neuronal Workspace theory, 177–78
neuroplasticity, 34, 155, 160–61
neuroscience, 16n12; dualism and, 17; *The Philosophical Foundations of Neuroscience*, 162; physicalism and, 9, 11
"Neuroscience and the Soul," Templeton Foundation and, 16n12

"A New Argument Against Materialism" (Plantinga), 8
*The New Directions in the Study of the Mind*, 9
NOD. *See* no-overdetermination
nomological impossibility, 79, 82–84
noncontradiction, law of, 79
nonphysical causes: with causal interaction between levels, 72n13; data and presence of, 71n8
nonreductive physicalism, 10, 53, 135n6
non sequitur, spatial relations, 125
Noordhof, Martin, 58
no-overdetermination (NOD), 52, 53, 56
"No Pairing Problem" (Bailey, Rasmussen and Horn), 8
no-report paradigm, 22, 26, 43n11, 44n21

Obama, Barack, 27, 135
*Objections of Physicalism* (Robinson, H.), 7
Occam's razor, 29
O'Connor, Timothy, 16n11, 16n13
Oderberg, David, 8, 87, 140, 141, 167n22
"On the Simplicity of the Soul" (Chisholm), 7
*On the Soul* (*De Anima*) (Aristotle), 5, 16n10, 101
ontology: Aristotelian powers, 142–45; extension with MBP, 150; human, 99–105; naturalism, 64; of substance versus aggregate, 90
organicism, 140, 153
organoids, brain, 36, 44n25
OUP. *See* Oxford University Press
overdetermination, 165n5; physical events and, 58; positing, 53–54
Owen, Adrian, 174
*The Oxford Handbook of Philosophy of Mind*, 74
Oxford University Press (OUP), 7, 8
oxygen atoms, brain and, 60

pain: brain and, 153; C-fiber activation and, 20–21, 30, 33, 42n3, 43n13; Kim on, 136n12; qualia and, 124
pairing problem. *See* causal pairing problem
pairing relation, causal problem and, 74–75
Pan, Jiahui, 172
panpsychism, 2
Paolini Poaletti, Michele, 137n26
Papineau, David, 58, 69, 71n4, 82
Parmenides, 90
Pasnau, Robert, 107–8, 146–47
passive paradigm, with DOC, 175–76
passive powers, 143–45, 149–50, 155, 157, 163–64
Patterson, Dennis, 162–63
Paul, Ellen Frankel, 8
Paul, Jeffrey, 8
Penfield, Wilder, 9, 65–66, 71n10
*Personal Identity* (Paul, E. F., Miller and Paul, J.), 8
PET. *See* positron emission tomography
phantasms, 147, 159
Phi, consciousness and, 179–82
Philosophers Pyramid, 90
*Philosophia Christi* (journal), 8, 9
*The Philosophical Foundations of Neuroscience* (Bennett and Hacker), 162
*Philosophical Perspectives* (journal), 7
*The Philosophical Review*, 1
*Philosophical Studies* (journal), 9
philosophy: *The Cambridge Dictionary of Philosophy*, 4; metaphysics and, 88; "A Return to Form in the Philosophy of Mind," 6; speculative, 1; *Stanford Encyclopedia of Philosophy*, 17
physical: connotations, 58–59; impossibility, 79–80; macrophysical and, 60–61; microphysical and, 59, 61; RPP, 152, 156, 158; supervenience and, 103

physical causal closure, 54, 62, 63
physical events: causes of, 67–68; mental and, 47–48, 49, 51–53, 55, 56, 61, 69; overdetermination and, 58
physicalism: *Christian Physicalism*, 10; consciousness and, 169; COP and, 56; dualism and, 5–6, 7, 8, 18; framework, 2; materialism and, 1, 4, 8, 9; neuroscience and, 9, 11. *See also* causal closure principle
*Physicalism and Its Discontents* (Gillett and Loewer), 8
*Physicalism, Or Something Near Enough* (Kim), 10–11, 73, 78, 123
*Physics* (Aristotle), 133, 143, 144
pig brains, 187
Plantinga, Alvin, 8, 69, 72n14
Plato: dualism, 100; with soul and body, 127; substance dualism and, 108, 109
Polonsky, Alex, 24–25
Popper, Karl, 41
positron emission tomography (PET), 23
*post hoc, ergo propter hoc*, 27
pot example, heating of metal, 143
powers: active, 143–45, 148–50, 163–64; Aristotelian ontology, 142–45; bodily/body, 193n19; defined, 142–43; IPP, 145; "Mental Powers," 167n31; mind-body, 148–53; passive, 143–45, 149–50, 155, 157, 163–64; soul, 145, 151; of teachers to teach, 150. *See also* Mind-Body Powers, and full NCC (neural correlates of consciousness)
presuppositions, conscious mind, 14–15
"The Primacy of the Mental" (Rickabaugh), 9
prime matter, 95
property dualism, 8–9, 12–13, 15, 34, 53, 104, 108–9
Pruss, Alexander, 72n12
psychological synchronization example, 76–77, 127–28

psychophysical laws: with justification for setting aside problem, 49–51; lack of, 47–51, 70–71
Putnam, Hilary, 33, 167n28
Pythagoras, 90

qualia: defined, 6; identity theory's viability and knowledge of, 34–35; nonphysical, 124
quantum mechanics, Copenhagen interpretation of, 68
*Questions on the Soul* (Aquinas), 110n5

Rasmussen, Joshua, 8, 9, 81
rational belief, CCP and, 68, 70
rationality, humanity and, 146
rational thought, 146–47, 171
*Red Hot Chili Peppers*, 35, 181
*The Rediscovery of the Mind* (Searle), 1, 4
reductionism, 11, 54
Requisite Physical Properties (RPP), 152, 156, 158
resurrection, of Jesus of Nazareth, 70
"A Return to Form in the Philosophy of Mind" (Haldane), 6
Rickabaugh, Brandon, 9
Robb, David, 16n11, 16n13
Robb, James H., 110n5
Robinson, Daniel, 8
Robinson, Howard, 7–10, 112n23
Route CO, 54–55, 70
Route OD, 53–55
Routledge, 7, 9, 10, 44n31
RPP. *See* Requisite Physical Properties
running example, 148–51

Schnakers, Caroline, 172
Scholasticism, 88
science: Allen Institute for Brain Science, 43n8; Association for the Scientific Study of Consciousness, 18; inevitable success of, 64–66. *See also* neuroscience

Searle, John, 1, 4
*Sensations and Brain Processes* (Smart), 29, 30
sensory stimuli, consciousness and, 3
sensory system, 173, 175
simplicity argument: analysis, 37–41; data scope, 38–40; dualism, 30–31; dualism misrepresented, 40–41; theoretical virtues and, 37–38
singularity, spacetime, 66–68, 81–82
Smart, J. J. C., 29–30, 32, 33, 40, 43nn13–14
Smith, Quentin, 82
Smythies, J. R., 7
Socrates, 90, 91, 99
soul: body and, 93, 98–99, 101–3, 106–7, 111n11, 113, 117, 126–31, 134–35, 135n4, 140–41, 147, 149, 151–52, 154, 165, 183, 193n15; brain and, 41; human embryo and, 133–34, 137n25; hylomorphic responses and spatial, 117–19; immaterial, 77; incarnate, 105–9; itself, 107–9; powers, 145, 151; *Questions on the Soul*, 110n5; thought and, 147
soul-body unity, causal pairing problem and, 102
*The Soul Hypothesis* (Baker, M. C., and Goetz), 8
spacetime, singularity of, 66–68, 81–82
spatiality: causality and, 82; God and, 136n10; necessity of, 75–77
spatial relations: with assumptions, unstated, 121–22; with beliefs, inconsistent, 122–25; causal pairing and, 76–77, 114, 117–18, 121, 122–23, 124, 126; with Kim and case for premises 2 and 3, 119–21; non sequitur, 125; questioning necessity of, 119–25
spatial soul, 117–19
Stalnaker, Robert, 29, 31–32, 43n12
*Stanford Encyclopedia of Philosophy*, 17

strict laws, 47, 48, 50
Stump, Eleonore, 100, 102, 108–9, 111n11
Sturgeon, Scott, 52, 58, 59, 61, 71nn5–6
*Subjects of Experience* (Lowe), 8
substance: aggregate versus, 89–92; body as, 91; hylomorphic responses and single, 115–17; inseparable aspect of, 104, 111n17; internal unity of, 89, 92; ontology of, 90; substantial form and, 91, 93
substance dualism: causal pairing problem and, 8, 11, 12, 46, 126; CCP and, 46, 73; Descartes, Plato and, 108, 109; falsifying, 41; mental causation and, 45–46; supporters of, 8, 12. *See also* neo-Thomistic hylomorphism
substantial forms: accidental and, 92–93; en-forming relation and, 98; substance and, 91, 93
*Summa Theologiae* (Aquinas), 99, 100, 105–7, 111n16, 147, 151, 166n14
supernatural explanation, 110n7, 131, 132, 136n23
supervenience, 103
Swinburne, Richard, 7–10, 15, 72n14, 72n15
*Synthese* (journal), 10, 137n26

Taliaferro, Charles, 10, 72n14
teacher, power to teach, 150
Templeton Foundation, 9, 16n12
*Teorema* (journal), 9
theoretical virtues, simplicity argument and, 37–38
Thomism, 3, 88, 103. *See also* Aquinas, Thomas
thought: brain and, 146–47; experiment, 34–35, 181–82; *Knowledge, Thought, and the Case for Dualism*, 9; rational, 146–47, 171; soul and, 147
time, measuring, 176

Tononi, Giulio, 178
"Top-Down Causation and the Human Brain" (Ellis), 72n13
*Topoi* (journal), 167n31
"Toward a Neurobiological Approach to Consciousness" (Crick and Koch), 4
"Treatise on Human Nature" (Aquinas), 99, 100
"Treatise On the One God" (Aquinas), 131
*The Trinity* (Augustine), 111n15
Trump, Donald, 128
Twain, Mark, 31–32
Tye, Michael, 9

"Undefeated Dualism" (Bogardus), 9
University of Virginia Press, 7
unresponsive wakefulness syndrome (UWS), 15n5, 171

Van Horn, Luke, 81
van Inwagen, Peter, 9, 105
vegetative state (VS), 3, 15n5, 171

Walls, J., 72n14
*The Waning of Materialism* (Koons and Bealer), 8
*The Weight of Simplicity in the Construction and Assaying of Scientific Theories* (Bunge), 37
"What Is a Neural Correlate of Consciousness?" (Chalmers), 18
"What Is It Like to Be Human (Instead of a Bat)?" (Bonjour), 9
Wheatstone, Charles (Sir), 24
Won, Chinook, 167n20
Wright, N. T., 72n15

Zimmerman, Dean, 8, 15
zombie nervous system, 35–37

# About the Author

**Matthew Owen** is a philosopher who teaches at Yakima Valley College in Washington State. He is also an affiliate faculty member of the Center for Consciousness Science at the University of Michigan Medical School. After earning a PhD at the University of Birmingham, England, he served as the Elizabeth R. Koch Research Fellow for Tiny Blue Dot Consciousness Studies at Gonzaga University, under the supervision of Dr. Christof Koch (Allen Institute for Brain Science) and Dr. Brian Clayton (Gonzaga University). Matthew has published research articles in various journals, including *Synthese: An International Journal for Epistemology, Methodology and Philosophy of Science*; *Frontiers in Systems Neuroscience*; *Entropy*; *Topoi: An International Review of Philosophy*; and *TheoLogica: An International Journal for Philosophy of Religion and Philosophical Theology*. Matthew and his wife Aryn live with their daughter Emma and lab Anselm just outside Yakima, Washington, in the foothills of the Cascades, where they enjoy hiking and skiing. Aryn serves the community as a flight nurse with Airlift Northwest, a branch of UW Medicine.

www.ingramcontent.com/pod-product-compliance
Lightning Source LLC
Chambersburg PA
CBHW020116010526
44115CB00008B/850